VLSI ARTIFICIAL NEURAL NETWORKS ENGINEERING

VLSI ARTIFICIAL NEURAL NETWORKS ENGINEERING

EDITED BY

Mohamed I. Elmasry
VLSI Research Group, University of Waterloo

KLUWER ACADEMIC PUBLISHERS
Boston/London/Dordrecht

Distributors for North America:
Kluwer Academic Publishers
101 Philip Drive
Assinippi Park
Norwell, Massachusetts 02061 USA

Distributors for all other countries:
Kluwer Academic Publishers Group
Distribution Centre
Post Office Box 322
3300 AH Dordrecht, THE NETHERLANDS

Library of Congress Cataloging-in-Publication Data

VLSI artificial neural networks engineering / edited by Mohamed
  I. Elmasry.
    p. cm.
  Includes bibliographical references and index.
  ISBN 0-7923-9493-3
  1. Neural networks (Computer science) 2. Integrated circuits-
-Very large scale integration. I. Elmasry, Mohamed I., 1943- .
QA76.87.V57 1994
621.39'9--dc20   94-26768
                                                                    CIP

Copyright © 1994 by Kluwer Academic Publishers

All rights reserved. No part of this publication may be reproduced, stored in a retrieval system or transmitted in any form or by any means, mechanical, photo-copying, recording, or otherwise, without the prior written permission of the publisher, Kluwer Academic Publishers, 101 Philip Drive, Assinippi Park, Norwell, Massachusetts 02061

Printed on acid-free paper.

Printed in the United States of America

To

Elizabeth, Carmen, Samir, Nadia and Hassan Elmasry

CONTENTS

1 AN OVERVIEW
Waleed Fakhr and Mohamed I. Elmasry **1**
   1.1 Introduction   1
   1.2 Biological Neural Networks   1
   1.3 Artificial Neural Networks (ANNs)   4
   1.4 Artificial Neural Network Algorithms   4
   1.5 Supervised Neural Networks   5
   1.6 Unsupervised Neural Networks   17
   1.7 Neural Network Architectures and Implementations   19
   1.8 Book Overview   22

2 A SAMPLED-DATA CMOS VLSI IMPLEMENTATION OF A MULTI-CHARACTER ANN RECOGNITION SYSTEM
Sameh E. Rehan and Mohamed I. Elmasry **33**
   2.1 Introduction   33
   2.2 ANN Implementation Techniques   34
   2.3 Developed CMOS Circuits for ANNs   46
   2.4 The Prototype MLP ANN Model Architecture   60
   2.5 MLP ANN Model Simulations   62
   2.6 ANN Circuit Simulations   72
   2.7 The Developed VLSI Architectures   72
   2.8 The Developed Two-Character ANN Recognizer   77
   2.9 The Proposed Multi-Character ANN Recognition System   79
   2.10 Conclusions   81

3 A DESIGN AUTOMATION ENVIRONMENT FOR MIXED ANALOG/DIGITAL ANNS
Arun Achyuthan and Mohamed I. Elmasry   **91**

    3.1   Introduction   91
    3.2   Mixed Analog/Digital ANN Hardware   93
    3.3   Overview of the Design Automation Environment   97
    3.4   Data Flow Graph   101
    3.5   The Analyzer   102
    3.6   The Design Library   112
    3.7   The Synthesizer   114
    3.8   Design Examples   122
    3.9   Conclusions   132

4 A COMPACT VLSI IMPLEMENTATION OF NEURAL NETWORKS
Liang-Yong Song, Anthony Vannelli and Mohamed I. Elmasry   **139**

    4.1   Introduction   139
    4.2   The Building Blocks   140
    4.3   The Circuit Implementation Example   148
    4.4   Expanding the Network with Multiple Chips   153
    4.5   Conclusions   155

5 AN ALL-DIGITAL VLSI ANN
Brian White and Mohamed I. Elmasry   **157**

    5.1   Introduction   157
    5.2   Neocognitron Neural Network Model   160
    5.3   Digi-Neocognitron (DNC): A Digital Neural Network Model for VLSI   164
    5.4   Character Recognition Example   177
    5.5   Advantages for VLSI Implementation   183
    5.6   Conclusions   185

6 A NEURAL PREDICTIVE HIDDEN MARKOV MODEL ARCHITECTURE FOR SPEECH AND SPEAKER RECOGNITION
Khaled Hassanein, Li Deng and Mohamed I. Elmasry   **191**

6.1	Introduction	191
6.2	Automatic Speech Recognition Methodologies	192
6.3	An ANN Architecture for Predictive HMMs	203
6.4	Discriminative Training of the Neural Predictive HMM	223
6.5	Speaker Recognition Using the Neural Predictive HMM	231
6.6	Conclusions	238

7 MINIMUM COMPLEXITY NEURAL NETWORKS FOR CLASSIFICATION
Waleed Fakhr, Mohamed Kamel and Mohamed I. Elmasry **247**

7.1	Introduction	247
7.2	Adaptive Probabilistic Neural Networks: APNN and ANNC	251
7.3	Bayesian PDF Model Selection	255
7.4	Maximum Likelihood and Maximum Mutual Information Training	260
7.5	Experimental Results	263
7.6	The Adaptive Feature Extraction Nearest Neighbor Classifier "AFNN"	268
7.7	Conclusions	279

8 A PARALLEL ANN ARCHITECTURE FOR FUZZY CLUSTERING
Dapeng Zhang, Mohamed Kamel and Mohamed I. Elmasry **283**

8.1	Introduction	283
8.2	Fuzzy Clustering and Neural Networks	284
8.3	Fuzzy Competitive Learning Algorithm	288
8.4	Mapping Algorithm Onto Architecture	291
8.5	Fuzzy Clustering Neural Network (FCNN) Architecture: Processing Cells	292
8.6	Comparison With The Fuzzy C-Mean (FCM) Algorithm	295
8.7	Conclusions	299

9 A PIPELINED ANN ARCHITECTURE FOR SPEECH RECOGNITION
Dapeng Zhang, Li Deng and Mohamed I. Elmasry **303**

9.1	Introduction	303

9.2	Definition and Notation	304
9.3	PNN Architecture: Processing Stages	306
9.4	Case Studies	312
9.5	Performance Analysis	316
9.6	Conclusions	320

INDEX   **323**

CONTRIBUTORS

Arun Achyuthan
Department of Electrical and Computer Engineering
University of Waterloo
Waterloo, Ontario, Canada

Li Deng
Department of Electrical and Computer Engineering
University of Waterloo
Waterloo, Ontario, Canada

Mohamed I. Elmasry
Department of Electrical and Computer Engineering
University of Waterloo
Waterloo, Ontario, Canada

Waleed Fakhr
Department of Electrical and Computer Engineering
University of Waterloo
Waterloo, Ontario, Canada

Khaled Hassanein
Department of Electrical and Computer Engineering
University of Waterloo
Waterloo, Ontario, Canada

Mohamed Kamel
Systems Design Engineering
University of Waterloo
Waterloo, Ontario, Canada

Sameh E. Rehan
Department of Electrical and Computer Engineering
University of Waterloo
Waterloo, Ontario, Canada

Liang-Yong Song
Department of Electrical and Computer Engineering
University of Waterloo
Waterloo, Ontario, Canada

Anthony Vannelli
Department of Electrical and Computer Engineering
University of Waterloo
Waterloo, Ontario, Canada

Brian White
Department of Electrical and Computer Engineering
University of Waterloo
Waterloo, Ontario, Canada

Dapeng Zhang
Department of Electrical and Computer Engineering
University of Waterloo
Waterloo, Ontario, Canada

Preface

Engineers have long been fascinated by how efficient and how fast biological neural networks are capable of performing such complex tasks as recognition. Such networks are capable of recognizing input data from *any of the five senses* with the necessary accuracy and speed to allow living creatures to survive. Machines which perform such complex tasks as recognition, with similar accuracy and speed, were difficult to implement until the technological advances of VLSI circuits and systems in the late 1980's. Since then, the field of VLSI Artificial Neural Networks (ANNs) have witnessed an exponential growth and a new engineering discipline was born. Today, many engineering curriculums have included a course or more on the subject at the graduate or senior undergraduate levels.

Since the pioneering book by Carver Mead; "Analog VLSI and Neural Systems", Addison-Wesley, 1989; there were a number of excellent text and reference books on the subject, each dealing with one or two topics. This book attempts to present an integrated approach of a single research team to VLSI ANNs Engineering.

Chapter 1 serves as an introduction. Chapters 2, 3, 4 and 5 deal with VLSI circuit design techniques (analog, digital and sampled data) and automated VLSI design environment for ANNs. Chapter 2 reports on a sampled data approach to the implementation of ANNs with application to character recognition. Chapter 2 also contains an overview of the different approaches of VLSI implementation of ANNs; explaining the advantage and disadvantage of each approach. In Chapter 3, the topic of design exploration of mixed analog/digital ANNs at the high level of the design hierarchy is addressed. The need for creating such a design automation environment, with its supporting CAD tools, is a necessary condition for the widespread use of application-specific chips for ANN implementation. In Chapter 4, the same topic of design exploration is discussed, but at the low level of the hierarchy and targeting analog implementation. Chapter 5 reports on an all-digital implementation of ANNs using the Neocognitron as the ANN model.

Chapters 6, 7, 8 and 9 deal with the application of ANNs to a number of fields. Chapter 6 addresses the topic of automatic speech recognition using neural predictive hidden Markov Models. Chapter 7 deals with the topic of classification using minimum complexity ANNs. Chapter 8 addresses the topic of pattern recognition using a fuzzy clustering ANNs. Finally, Chapter 9 deals with speech recognition using pipelined ANNs.

It is my hope that this book will contribute to our understanding of this new and exciting discipline; VLSI ANNs Engineering.

M.I. Elmasry
Waterloo, Ontario
Canada

Acknowledgements

I would like to first acknowledge the countless blessings of God Almighty throughout my life. This book, in particular, has enhanced my belief in the Creator and has exposed me to some of His magnificent Creation as related to biological neural networks.

My graduate students and research associates were excellent companions on a difficult road towards understanding VLSI ANNs Engineering. Many have contributed to this book and to them I am very grateful. Special thanks are due to Drs. W. Fakhr, K. Hassanein and D. Zhang for their help in the preparation of this book. The financial support of the granting agency NSERC is appreciated.

1
AN OVERVIEW
Waleed Fakhr and Mohamed I. Elmasry

1.1 INTRODUCTION

Artificial Neural Networks have become an increasingly popular field of research in many branches of science. These include computer engineering and computer science, signal processing, information theory, and physics. Due to the variety of fields of interest and applications, a broad range of artificial neural network models and VLSI implementations has emerged. In all these fields, the term "neural networks" is characterized by a combination of adaptive learning algorithms and parallel distributed implementations [1]. Although artificial neural networks are biologically motivated, their resemblances to the brain models are not straightforward. Nevertheless, recent research in both artificial and biological neural networks have benefited one another. In order to appreciate the differences and similarities between both, we start by a brief introduction to biological neural networks.

1.2 BIOLOGICAL NEURAL NETWORKS

It is interesting to compare the human brain with a serial modern Von Neuman computer from the information processing aspects as shown in Table 1.1 [1]. Although the neuron's switching time (a few milliseconds) is about a million times slower than modern computer elements, they have a thousand-fold greater connectivity than today's supercomputers. Furthermore, the brain is very power efficient - it consumes less than 100 watts- by contrast a supercomputer may dissipate 10^5 watts [1].

Table 1.1 Human Brain and Modern Computer. A Comparison

Human Brain	Modern Computer
10^{11} neurons and 10^{14} synapses	10^9 transistors
Analog	Digital
Fanout 10^3	Fanout 3
Massively Parallel	Largely Serial
Switching time 10^{-3} s	Switching time 10^{-9} s

Neurons and the interconnection synapses constitute the key elements for neural information processing. Most neurons possess tree-like structures called dendrites which receive incoming signals from other neurons across junctions called synapses [2]. Some neurons communicate with only a few nearby ones, whereas others make contact with thousands.

There are three parts in a neuron: (1) a neuron cell body, (2) branching extensions called dendrites for receiving input, and (3) an axon that carries the neuron's output to the dendrites of other neurons. The synapse represents the junction between an axon and a dendrite. How two or more neurons interact remains largely mysterious, and complexities of different neurons vary greatly. Generally speaking, a neuron sends its output to other neurons via its axon. An axon carries information through a series of action potentials, or waves of current, that depend on the neuron's voltage potential. More precisely, the membrane generates the action potential and propagates down the axon and its branches, where axonal insulators restore and amplify the signal as it propagates, until it arrives at a synaptic junction. This process is often modeled as a propagation rule represented by a net value $u(.)$, as in Figure 1.1.a.

A neuron collects signals at its synapses by summing all the excitatory and inhibitory influences acting upon it. If the excitatory influences are dominant, then the neuron fires and sends this message to other neurons via the outgoing synapses. In that sense, the neuron function may be modeled as a simple threshold function $f(.)$. As shown in Figure 1.1.a, the neuron fires if the combined signal strength exceeds a certain threshold.

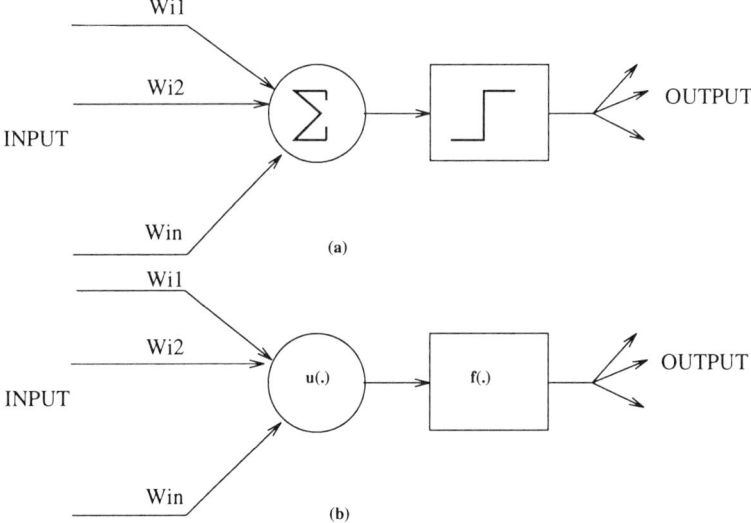

Figure 1.1 (a) A Simplified Neural Network Model with Linear Threshold Functions. (b) A General Neural Model.

1.3 ARTIFICIAL NEURAL NETWORKS (ANNS)

In general, neurons and axons are mathematically modeled by activation functions and net functions (or basis functions) respectively, as shown in Figure 1.1.b. Lacking more advanced knowledge on biological nervous systems, it is impossible to specifically define the neuron functionalities and connection structures merely from a biological perspective. Consequently, the selection of these functions often depends on the application the neural models are for, or the way they are going to be implemented. In other words, application-driven artificial neural networks are only loosely tied to the biological realities. They are, however, strongly associated with high-level and intelligent processing in recognition and classification [1]. They have the potential to offer a truly revolutionary technology for modern information processing. The strength of artificial neural networks hinges upon three main characteristics [1]:

Adaptiveness and self-organization: it offers robust and adaptive processing capabilities by adopting adaptive learning and self-organization rules.

Nonlinear network processing: it enhances the network's approximation, classification and noise-immunity capabilities.

Parallel processing: it usually employs a large number of processing cells enhanced by extensive interconnectivity.

These characteristics have played an important role in neural network's applications and implementations.

1.4 ARTIFICIAL NEURAL NETWORK ALGORITHMS

There is a large and diverse number of neural network algorithms, driven by different applications, and implemented by many different techniques. Figure 1.2 shows a taxonomy of the major neural network paradigms, which include most of the existing ones. These algorithms may be divided into two major categories according to their learning paradigms, supervised networks and unsupervised networks, where the fixed weights type is considered as a special subset of the supervised networks.

In the following sections we are going to discuss the above paradigms, so as to give the reader a brief description of their theory and fields of applications. For

An Overview

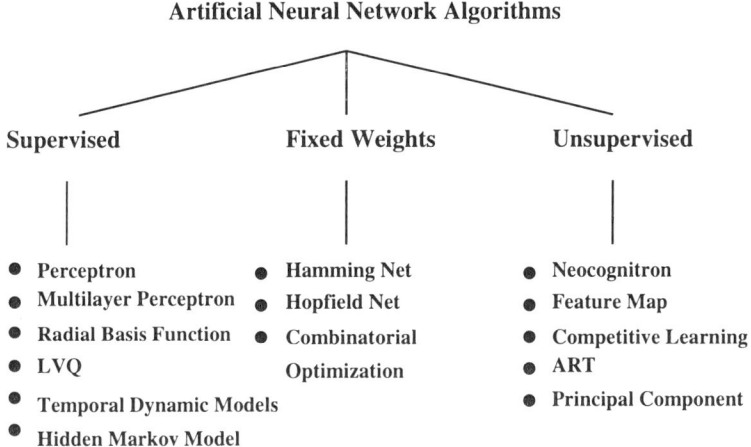

Figure 1.2 A Taxonomy of Neural Networks

a more detailed discussion, the reader is referred to [1], which also contains a large number of references on the field.

1.5 SUPERVISED NEURAL NETWORKS

Supervised neural networks may be divided into static and dynamic models. In static models, the network makes a decision or a mapping based only on its current input pattern. In dynamic models, the network makes the decision based on the current input and past inputs/outputs. We start the review by static models.

1.5.1 The Perceptron

The perceptron, shown in Figure 1.3, was developed by Rosenblatt in 1958 [3]. The input is an n-dimensional vector, and the perceptron forms a linear weighted sum of the input vector and adds a threshold value θ. The result is

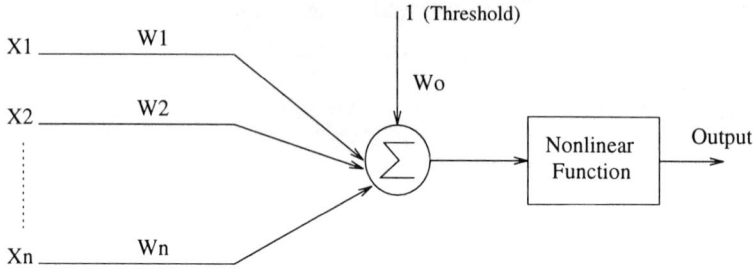

Figure 1.3 The Perceptron

then passed through a nonlinearity, which may be a hard limiter or a sigmoid. The sigmoid nonlinearity is popular because it is differentiable, and for other reasons as well. For example, many applications require a continuous-valued output rather than the binary output produced by the hard limiter. In addition, it is particularly well suited to pattern recognition applications because it produces an output between 0 and 1 that can often be interpreted as a probability estimate [4].

There are numerous learning algorithms for the perceptron. They include the perceptron learning algorithm [3], the least mean square learning algorithm "LMS" [5], and many others [6]. The LMS algorithm is a special case of the Back-propagation algorithm (for the multilayer perceptron), and will be discussed shortly.

1.5.2 The Multilayer Perceptron "MLP"

The capabilities of the single perceptron are limited to linear decision boundaries and simple logic functions. However, by cascading perceptrons in layers we can implement complex decision boundaries and arbitrary boolean expressions. The individual perceptrons in the network are called neurons or nodes, and differ from Rosenblatt's perceptron in that a sigmoid nonlinearity is commonly used in place of the hardlimiter. The input vector feeds into each of the second layer perceptrons, and so on, as shown in Figure 1.4. For classification problems, Lippmann demonstrated that a 2-layer MLP can implement arbitrary convex decision boundaries [7]. Later it was shown that a 2-layer network can

An Overview

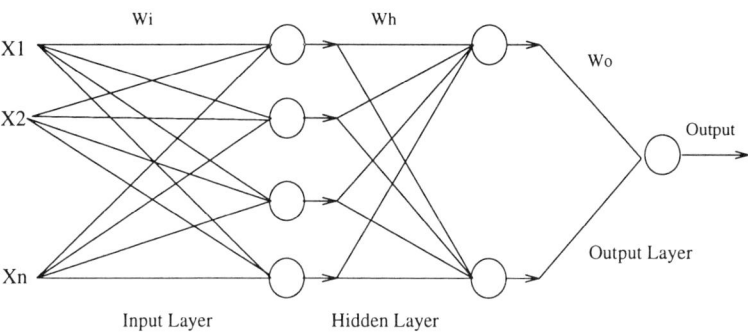

Figure 1.4 A Three-Layer Multi-Layer Perceptron (MLP)

form an arbitrarily close approximation to any continuous nonlinear mapping [8]. However, these results do not necessarily imply that there is no benefit to having more than two layers. For some problems, a small 3-layer network can be used where a 2-layer network would require an infinite number of nodes [9]. It has also been shown that there are problems which require an exponential number of nodes in a 2-layer network that can be implemented with a polynomial number of nodes in a 3-layer network [10]. None of these results require the use of the sigmoid nonlinearity in particular. The proofs assume only that the nonlinearity is a continuous smooth monotonically increasing function that is bounded above and below. In addition, none of the above results require that the nonlinearity be present at the output layer. Thus, it is quite common to use linear output nodes since this tends to make learning easier, and sometimes faster [11].

1.5.2.1 Back-propagation Learning Algorithm

One of the limitations of Rosenblatt's original formulation of the MLP was the lack of an adequate learning algorithm. Algorithms were eventually developed to overcome this limitation [12-14]. The most common approach is to use a gradient descent algorithm, but the key difficulty in deriving such an algorithm for the MLP was that the gradient is zero almost everywhere when the hard-limiting nonlinearity is used. The nonlinearity must be present however, because without it, the MLP would implement nothing more than a linear transformation at each layer, in which case the MLP could be reduced to an equivalent single layer network. The solution is to use a nonlinearity that is differentiable. The nonlinearity most often used is the sigmoid function. With this nonlinearity, it is possible to implement a gradient search of the weight space, which is known as the Back-propagation algorithm.

In the MLP, training data is given, where each pattern x_n has a corresponding target output vector d_n. The learning is done by minimizing the mean square error between the desired output vector and the actual one, for all given patterns, where the minimized criterion is given by:

$$G = \Sigma_{n=1}^{N} G_n \tag{1.1}$$

where N is the number of training patterns, G_n is the total squared error for the n_{th} pattern:

$$G_n = \frac{1}{2} \Sigma_{q=1}^{Q} (y_q(x_n) - d_q(x_n))^2 \tag{1.2}$$

and Q is the number of nodes in the output layer. An arbitrary weight w in the MLP is determined iteratively according to:

$$w(k+1) = w(k) - \mu \frac{\delta G}{\delta w} \tag{1.3}$$

where μ is a positive constant called the learning rate. The implementation of this algorithm uses the chain rule of differentiation to find the gradients for input and hidden layers weights.

The weights are typically initialized to small random values. This starts the search in a relatively "safe" position. The learning rate can be chosen in a number of different ways. They can be the same for every weight in the network, different for each layer, different for each node, or different for each weight in the network. In general it is difficult to determine the best learning rate, but a useful rule of thumb is to make the learning rate for each node inversely proportional to the average magnitude of vectors feeding into the node. Several attempts have been made to adapt the learning rate as a function of the local curvature of the surface. The simplest approach, and one that works quite well in practice, is to add a *momentum* term, to each weight update. This term makes the current search direction an exponentially weighted average of past directions, and helps keep the weights moving across flat portions of the error surface after they have descended from the steep portions.

The process of computing the gradient and adjusting the weights is repeated until a minimum (or a point sufficiently close to the minimum) is found. In practice it may be difficult to automate the termination of the algorithm. However, there are several stopping criteria that can be considered. The first is based

An Overview

on the magnitude of the gradient. The algorithm can be terminated when this magnitude is sufficiently small, since by definition the gradient will be zero at the minimum. Second, one might consider stopping the algorithm when G falls below a fixed threshold. However this requires some knowledge of the minimal value of G, which is not always known. In pattern recognition problems, one might consider stopping as soon as all of the training data are correctly classified. This assumes, however, that the network can actually classify all of the training data correctly, which will not always be the case. Even if it can do so, this stopping criterion may not yield a solution that generalizes well to new data. Third, one might consider stopping when a fixed number of iterations have been performed, although there is little guarantee that this stopping condition will terminate the algorithm at a minimum. Finally, the method of cross-validation can be used to monitor generalization performance during learning, and terminate the algorithm when there is no longer an improvement. The method of cross-validation works by splitting the data into two sets: a training set which is used to train the network, and a test set which is used to measure the generalization performance of the network. During learning, the performance of the network on the training data will continue to improve, but its performance on the test data will only improve to a point, beyond which it will start to degrade. It is at this point, where the network starts to overfit the training data, that the learning algorithm is terminated.

1.5.2.2 MLP Issues and Limitations

The MLP is capable of approximating arbitrary nonlinear mappings, and given a set of examples, the Back-propagation algorithm can be called upon to learn the mapping at the example points. However, there are a number of practical concerns. The first is the matter of choosing the network size. The second is the time complexity of learning. That is, we may ask if it is possible to learn the desired mapping in a reasonable amount of time. Finally, we are concerned with the ability of our network to generalize; that is, its ability to produce accurate results on new samples outside the training set.

First, theoretical results indicate that the MLP is capable of forming arbitrarily close approximations to any continuous mapping if its size grows arbitrarily large [15]. In general, it is not known what finite size network works best for a given problem. Further, it is not likely that this issue will be resolved in the general case since each problem will demand different capabilities from the network. Choosing the proper network size is important. If the network is too small, it will not be capable of forming a good model of the problem.

On the other hand, if the network is too big then it may be too capable. That is, it may be able to implement numerous solutions that are consistent with the training data, but most of these are likely to be poor approximations of the actual problem. Ultimately, we would like to find a network whose size best matches the underlying problem, or captures the structure of the data. This issue will be discussed in more detail in chapter 7.

With little or no knowledge of the problem one must determine the network size by trial and error. One approach is to start with the smallest possible network and gradually increase the size until the performance begins to level off. A closely related approach is to "grow" the network. The idea here is to start with one node and create additional nodes as they are needed. Approaches that use such a technique include Cascade Correlation [16], the Group Method of Data Handling "GMDH" [17], Projection Pursuit [18,19], and others [20]. Another possibility is to start with a large network and then apply a pruning technique that destroys weights and/or nodes which end up contributing little or nothing to the solution [21,22].

In summary, the MLPs are good at both classification and nonlinear regression, as long as their architectures are selected carefully.

1.5.3 Radial Basis Function Networks

A Radial Basis Function "RBF" network [23], as shown in Figure 1.5, is a two-layer network whose output nodes form a linear combination of the basis (or kernel) functions computed by the hidden layer nodes. The basis functions in the hidden layer produce a localized response to input stimulus. That is, they produce a significant nonzero response only when the input falls within a small localized region of the input space. For this reason this network is sometimes referred to as the localized receptive field network [24,25]. Although implementations vary, the most common basis is a Gaussian kernel function of the form:

$$u_j = \exp - \frac{1}{2\sigma_j^2} \Sigma_{i=1}^{I} (x_i - m_{ij})^2 \qquad j = 1, 2, ..., K \qquad (1.4)$$

where u_j is the output of the j_{th} node in the first layer, x_i is the i_{th} component of the input pattern, m_{ij} is the i_{th} component of the j_{th} node in the first layer, i.e., the center of the Gaussian node j, σ_j^2 is the normalization parameter for the j_{th} node, and K is the number of nodes in the first layer. The node outputs are in the range from zero to one so that the closer the input is to the center of the Gaussian, the larger the response of the node.

An Overview

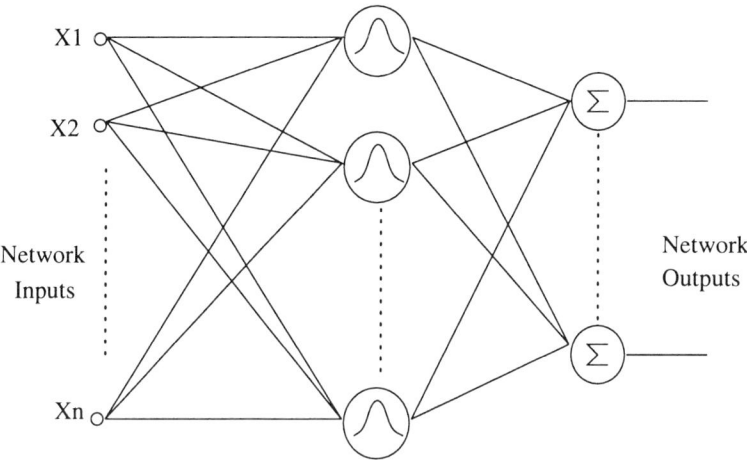

Figure 1.5 The Radial Basis Function Network (RBF)

The name radial basis function comes from the fact that these Gaussian kernels are radially symmetric; that is, each node produces an identical output for inputs that lie a fixed radial distance from the center of the j_{th} kernel.

The output layer node equations are given by:

$$y_l = \Sigma_{j=1}^{K} w_{jl} u_j \qquad l = 1, 2, ..., L \qquad (1.5)$$

where y_l is the output of the l_{th} node, and w_{jl} is the weight connecting the j_{th} hidden node output to the l_{th} output node.

The RBF network can be used for both classification and functional approximation, just like the MLP. In theory, the RBF network, like the MLP, is capable of forming an arbitrary close approximation to any continuous nonlinear mapping [26]. The primary difference between the two is in the nature of their basis functions. The hidden layer nodes in an MLP form sigmoidal basis functions which are nonzero over an infinitely large region of the input space, while the basis functions in the RBF network cover only small localized regions. While some problems can be solved more efficiently with sigmoidal basis functions, others are more amenable to localized basis functions.

1.5.3.1 RBF Learning Algorithm

There are a variety of approaches to learning in the RBF network. Most of them start by breaking the problem into two stages: learning in the hidden layer, followed by learning in the output layer [24,25]. Learning in the hidden layer is typically performed using an unsupervised method, i.e., a clustering algorithm, while learning in the output layer is supervised. Once an initial solution is found using this approach, a supervised learning algorithm is sometimes applied to both layers simultaneously to fine-tune the parameters of the network. There are numerous clustering algorithms that can be used in the hidden layer. A popular choice is the K-means algorithm [6,27], which is perhaps the most widely known clustering algorithm because of its simplicity and its ability to produce good results.

The normalization parameters, σ_j^2, are obtained once the clustering algorithm is complete. They represent a measure of the spread of the data associated with each node. The most common way is to make them equal to the average distance between the cluster centers and the training patterns.

Learning in the output layer is performed after the parameters of the basis functions have been determined; that is, after learning in the first layer is complete. The output layer is typically trained using the least mean square "LMS" algorithm [5].

1.5.4 Other Pattern Classifiers

Thus far we have focused primarily on the Multilayer Perceptron and the Radial Basis Function networks. While these are not the only neural network pattern classifiers, we feel that they are arguably the most popular and most clearly illustrate the major features of neural network approaches to supervised, static pattern classification. In this section, we briefly discuss some other pattern classifiers. These include more traditional techniques such as the Gaussian classifier [28], Gaussian mixture methods [6,28,29], Parzen windows, including the probabilistic neural network "PNN" [6,28], polynomial classifiers [6,28], nearest-neighbor techniques, including the learning vector quantization "LVQ" [28,30,31], and tree-based methods [32].

The Gaussian classifier is a consequence of applying Bayes decision rule for the case where the probability density functions for each class are assumed to be Gaussian [28]. To implement this classifier, one need only to estimate the

An Overview 13

mean vector and covariance matrix for each class of data, and substitute these estimates into the decision rule. The decision rule can be simplified to yield a polynomial classifier which is in general quadratic, but reduces to linear in the case where the covariance matrices for both classes are equal. There are numerous ways of implementing the Gaussian classifier. The following three approaches are important because they form the foundation for the other classifier schemes discussed in this section. The direct approach uses estimates of the class density functions which are substituted directly (without simplification) into the Bayes rule. The discriminant function approach uses a simplified decision function which, as mentioned above, is a polynomial which is at most quadratic. The distance classifier approach uses a decision rule which assigns input patterns to the "closest class". In the Gaussian case, the closest class is determined by the distance from the input pattern to the class means. In the Gaussian problem, all of these methods yield equivalent results, and all are optimal, however they lead to different results when the problem is not a Gaussian one. These three basic approaches can be extended by:

(1) Using more elaborate density function estimates,

(2) Using more powerful polynomial discriminant functions, or

(3) Using a more general distance classifier.

More elaborate density estimates can be formed using a Gaussian mixture [6,28,29]. In this case the density function for each pattern class is approximated as a mixture of Gaussian density functions. Determining the optimal number of Gaussians in the mixture, and their parameters can prove to be a difficult problem, which is also discussed in chapter 7. Clustering algorithms like the K-means algorithm can be used to determine the mean vectors of these Gaussians. Covariance matrices can then be determined for each cluster once the mean vectors are estimated. This approach has a similarity to the RBF network. The RBF network can be viewed as a weighted Gaussian mixture, where the weights of the Gaussians are determined by the supervised learning algorithm at the output layer. In theory, this approach is capable of forming arbitrarily complex decision boundaries for classification. An alternative approach is to place a window function, e.g., a Gaussian window, at every training sample. This type of approach is usually referred to as the Parzen window method. While the positions of the windows are defined automatically; determining the window widths can be a difficult task. Many neural network models are related to this approach. These models include the probabilistic neural network "PNN" which uses a Gaussian window, and the CMAC model which uses a rectangular window [33].

The above methods are often referred to as memory-based models since they represent generalizations of methods that work by memorizing the response to the training data. The usefulness of these techniques is generally determined by how efficiently they can cover the relevant portions of the input space.

The second of the three basic approaches mentioned above is the discriminant function approach. Perhaps the simplest discriminant function is a linear discriminant. This gives rise to a decision boundary between the classes that is linear, i.e., a hyperplane.

Linear discriminants are important, not only because they are practical, but also because they have been shown to be optimal for many problems. The perceptron is a simple neural network classifier which employs a linear discriminant function. The MLP can be viewed as an extension of the discriminant function approach which is capable of forming arbitrarily complex decision boundaries.

Tree-based classifiers can also be viewed as a nonlinear discriminant function approach. In a tree-based approach, the classifier is constructed by a series of simple greedy splits of the data into subgroups. Each subgroup is then split recursively, so that the resulting classifier has a hierarchical binary tree structure.

Typically, each greedy split finds the best individual component along which to split the data, although the algorithm can apply a more general split such as a linear or polynomial discriminant function at each node in the tree. Classification assignments are made at the leaves of the tree. The resulting decision boundary is generally a piece-wise linear boundary. This approach is particularly useful in problems where the input patterns contain a mixture of symbolic and numerical data. It also provides a rule-based interpretation of the classification method (the decisions made at each tree node determine the rules). Many techniques for growing and pruning tree classifiers have also been proposed.

The third approach is the distance classifier. The simplest distance classifier corresponds to the case of the Gaussian classifier with equal covariance matrices. In this case, the Euclidean distance from the input pattern to the mean of each class is used to make the classification decision. When the covariance matrices for different classes are not equal, the Euclidean distance becomes a Mahalanobis distance, and the resulting decision boundary is quadratic. Extensions of the distance classifier include the k-nearest neighbor "k-NN" and the learning vector quantization "LVQ" methods. The k-NN approach computes the distance between the input pattern and a set of labeled patterns, keeping track of the k-set of closest patterns from the labeled set. The input is then assigned to the class with most members in the k-set. The labeled set is formed directly from the training data. In fact, in the standard k-NN classifier, all training data are used in the labeled set. However, in the interest of reducing the computational and storage requirements, algorithms have been developed to reduce the size of the labeled set. The decision boundaries formed by this method are piecewise linear.

The LVQ method works exactly like a 1-NN classifier, except that the set of labeled patterns is formed differently. This set is typically obtained by clustering the training data (to reduce the number of labeled patterns), and then using a supervised learning algorithm to move cluster centers into positions that reduce the classification error.

Determining which classification method works best in a given application usually involves some degree of trial and error. Generally speaking, most of the approaches mentioned above can be designed to yield near-optimal classification performance to most problems. The real difference between them lies in other areas such as their time complexity of learning, their computational and storage requirements, their suitability for VLSI implementations, their generalization capability, and their potential use as estimators of a posteriori probabilities.

1.5.5 Nonlinear Temporal Dynamic Models

So far we have discussed supervised neural networks which make their decisions based on static patterns. Many real-world applications must involve nonlinear temporal dynamic models, which can be divided into non-recurrent time-delay neural networks "TDNNs", and recurrent neural networks "RNNs".

1.5.5.1 Time Delay Neural Network "TDNN"

Before we discuss neural networks that are truly dynamic, consider how an MLP is often used to process time series data. It is possible to use a static network to process time series data by simply converting the temporal sequence into a static pattern by unfolding the sequence over time. That is, time is treated as another dimension in the problem. From a practical point of view we can only afford to unfold the sequence over a finite period of time.

This can be accomplished by feeding the input sequence into a tapped delay line of finite extent, then feeding the taps from the delay line into a static neural network architecture like the MLP. An architecture like this is often referred to as a Time Delay Neural Network "TDNN" [34]. It is capable of modeling systems where the output has a finite temporal dependence on the input, similar to the finite impulse response "FIR" filter. Because there is no feedback in this network, it can be trained using the standard Back-propagation algorithm.

The TDNN has been used quite successfully in many applications. The NETtalk project [35] used the TDNN for text-to-speech conversion. In this system, the input consisted of a local encoding of the alphabet and a small number of punctuation symbols. The output of the network was trained to give the appropriate articulatory parameters to a commercial speech synthesizer. These signals represented the phoneme to be uttered at the point of text corresponding to the center character in the tapped-delay line. A version of the TDNN with weight sharing and local connections has been used for speech recognition with excellent results [36]. The TDNN has also been applied to nonlinear time series prediction problems. The same approach using RBF networks was also investigated [25].

1.5.5.2 Recurrent Neural Networks

A major alternative to TDNN is to incorporate delay feedback into temporal dynamic models, making them recurrent [13,37,38]. For example, an MLP may be made recurrent by introducing time-delay loops to the input, hidden, and/or output layers. Another way is to route delay connections from one layer to another. As a result of such a structural change, the gradient computation for the recurrent neural networks involves a complex Back-propagation rule through both time and space. Recurrent neural networks may have output feedback or state feedback, e.g., the Hopfield network. Their applications vary from solving optimization problems, time-series prediction, nonlinear identification and control, and associative memories.

All the above temporal dynamic networks are called deterministic, where the memory of the system can be represented by time-delay units. In stochastic networks, such as the hidden Markov models, the memory mechanism is implicitly manifested by a state transition matrix, which is trained to best model the temporal behavior. This model will be discussed in more details in chapter 6.

An Overview 17

1.6 UNSUPERVISED NEURAL NETWORKS

Unsupervised neural networks may be divided into two broad categories; those which perform clustering, and those which perform feature extraction, e.g., principal component analysis. The competitive learning networks belong to the first category, and include the Self-Organization Networks [39], the Adaptive Resonance Theory "ART" [40], and the Neocognitron [41]. Also, models which perform data clustering by probability density estimation [42] belong to the first category. The second category includes models which iteratively extract linear or nonlinear features from the unlabeled data, e.g., principal component extraction [43,44].

1.6.1 Competitive Learning Networks

Using no supervision from any teacher, unsupervised networks attempt to adapt the weights and verify the results based only on the input patterns. One popular scheme for such adaptation is the competitive learning rule, which allows the units to compete for the exclusive right to respond to (i.e., to be trained by) a particular training pattern. It can be viewed as a sophisticated clustering technique, whose objective is to divide a set of input patterns into a number of clusters such that the patterns of the same cluster exhibit a certain degree of similarity. Various kinds of competitive learning networks have been developed. The training rules are often the Hebbian rule for the feed-forward network, and the winner-take-all rule for the lateral network. They have the following distinctive features:

(1) Competitive learning networks extend the simple Hebbian rule to the very sophisticated competition-based rule. The training procedure is influenced by both the Hebbian rule and the winner-take-all rule.

(2) In order to implement competitive learning, the lateral networks are usually inhibitory. A winner-take-all circuit is used to select a winner, based on a distance metric over the pattern vector space. More important, a unit learns if and only if it wins the competition among all the other units.

In practical unsupervised learning, the number of classes may be unknown a priori. Therefore, the number of output nodes cannot be accurately determined in advance. To overcome this difficulty, it is useful to introduce a mechanism allowing adaptive expansion of the output layer until an adequate size is reached. To this end, two well-known techniques may be adopted: the vector quantizer "VQ" and the adaptive resonance theory "ART".

1.6.1.1 Vector Quantizer "VQ"

A vector quantizer is a system for mapping a set of vectors into a finite cluster for digital storage or communication. It is one of the very first unsupervised learning techniques. From the coding perspective, the goal of quantization is to obtain the best possible fidelity for the given compression rate. A vector quantizer maps a set of discrete vectors into a representation suitable for communication over a digital channel. In many applications, such as digital communications, better data compression has been achieved by using vector quantizers instead of scalar ones. The VQ procedure is summarized below:

(1) Given a new pattern, identify the best old cluster to admit the pattern.

(2) The centroid of the selected cluster is adjusted to accommodate its new member.

(3) If none of the old clusters can admit the new pattern, a new cluster will be created for it.

(4) Repeat the procedure for all the successive patterns.

1.6.1.2 Adaptive Resonance Theory "ART"

More recently, a more sophisticated clustering technique for adaptively adjusting the number of clusters was introduced by Carpenter and Grossberg. This is called adaptive resonance theory "ART". In addition to the forward networks between the input and output neurons, a backward network is adopted for vigilance test. Two versions of ART have been developed: ART1 for binary-valued patterns and ART2 for continuous-valued patterns. A new neuron can be created in ART for an incoming input pattern if it is determined (by a vigilance test) to be sufficiently different from the existing clusters. Such a vigilance test is incorporated into the adaptive backward network. Similarities between the ART framework and adaptive k-means algorithm have been recently discussed in [45].

1.6.2 The Self-Organizing Feature Map

Suppose that an input pattern has N features and is represented by a vector x in an N-dimensional pattern space. The network maps the input patterns to an output space. The output space in this case is assumed to be one-dimensional or two-dimensional arrays of nodes, which possess a certain topological order. The question is how to train the network so that the ordered relationship can be preserved. Kohonen proposed allowing the output nodes to interact laterally, leading to the self-organizing feature map.

The most prominent feature in the self-organizing map is the concept of excitatory learning within a neighborhood around the winning neuron. The size of the neighborhood slowly decreases with each iteration. A more detailed description of the learning phase is provided in [39]. In the retrieving phase, all the output neurons compute the Euclidean distance between the weights and the input vector and the winning neuron is the one with the shortest distance.

1.6.3 Neocognitron: Hierarchically Structured Model

Most existing neural network models could not effectively deal with patterns that were shifted in position or distorted. To combat this problem, Fukushima and Miyaki [46] proposed the so-called neocognitron model, which is especially applicable to space image recognition regardless of position and distortion.

1.7 NEURAL NETWORK ARCHITECTURES AND IMPLEMENTATIONS

Most neural network models are extremely demanding in both computation and storage requirements. An enormous amount of computation has to be spent on training the networks. In the retrieving phase, extremely high throughputs are required for real time processing. The attractiveness of the real-time processing hinges upon its massively parallel processing capability [47]. In this regard, there are two fundamental issues [1]:

1.7.1 Parallelism of Neural Algorithms

A thorough theoretical understanding of explicit and inherent parallelism in neural models can help design cost-effective real-time processing hardware. Most neural models, such as the MLP with Back-propagation network are very parallelizable. Some models are inherently suitable to parallel processing, for example, hierarchical perceptron and hidden Markov models. Yet some other neural models, for example, the original Hopfield model, Boltzmann machine, and annealing techniques can become parallelizable after proper modification of the original models.

1.7.2 Parallelism of Neural Architectures

Most neural algorithms involve primarily those operations that are repetitive and regular. They can be efficiently mapped to parallel architectures. For these classes of algorithms, an attractive and cost-effective architecture choice is an array processor, which uses mostly local interconnections. This paves the way for massively parallel processing, which represents the most viable future solution to real-time neural information processing.

Many ANN models require dedicated neural processing circuits, which are aimed at high performance for specific applications. Due to the maturity of design tools and the Very Large-Scale Integration (VLSI) technology, CMOS circuits are now widely adopted in many analog, digital, and mixed signal ANN design.

1.7.3 Analog versus Digital Design [1]

Both analog and digital techniques have demonstrated a high degree of success in their own application domains. The selection between digital and analog circuits depends on many factors, for example, speed, precision, adaptiveness, programmability, and transfer/storage of signals.

1.7.3.1 Analog Implementation of ANNs

In dedicated analog devices, a neuron is basically a differential amplifier with synaptic weights implemented by resistors. Thus, many neurons can fit on a single chip. Analog circuits can process more than 1 bit per transistor with a very high speed. The asynchronous updating property of analog devices offers qualitatively different computations from those offered by digital devices. For real-time early vision processing, dedicated analog processing chips offer arguably the most appealing alternative. For example, analog circuits offer inherent advantages on (1) the computation of the sum of weighted inputs by current or charge packets, and (2) the nonlinear effects of the devices facilitating realization of sigmoid-type functions. Because of the vital integration between analog sensors and information preprocessing/postprocessing, analog circuits will continue to have a major impact on dedicated neuron processing designs.

Although analog circuits are more attractive for biological-type neural networks, their suitability for connectionist-type networks remains very questionable. For example, compared with digital circuits, analog circuits are more susceptible to noise, crosstalk, temperature effects, and power supply variations. Although nonvolatile storage of analog weights provides high synaptic density, they are not easily programmable. In fact, the higher the precision, the more chip area is required. Thus, analog precision is usually limited to no more than 8 bits. In resistor-capacitor circuitry, low current consumption calls for high-resistance resistors. In switch-capacitor and resistor-capacitor circuitry, the low-noise constraint limits the minimal transistor surfaces and capacitors. In short, the combined factors of precision, noise, and current consumption lead to a larger chip area.

1.7.3.2 Digital Implementation of ANNs

For connectionist networks, digital technology offers very desirable features such as design flexibility, learning, expandable size, and accuracy. Digital designs have an overall advantage in terms of system-level performance. Dynamic range and precision are critical for many complex connectionist models. Digital implementation offers much greater flexibility of precision than its analog counterpart. Design of digital VLSI circuits is supported by mature and powerful CAD technology, as well as convenient building-block modular design. Digital circuits also have advantage in access to commercial design software and fast turnaround silicon fabrication. The disadvantages of digital circuits are, for example, bulky chip areas and (sometimes) relatively slow speeds.

A detailed discussion of analog versus digital VLSI implementation of ANNs is covered in chapter 2, while in chapters 3 and 4 ANN, VLSI design methodologies are proposed.

1.8 BOOK OVERVIEW

1.8.1 Chapter 2: Sampled-Data CMOS VLSI Implementation for ANN Character Recognition System

Dedicated VLSI circuits, which provide a compact and fast implementation of artificial neural networks (ANNs), can release the full power of these structures. A mixed-mode analog/digital VLSI implementation of ANNs offers a tradeoff solution for speed, area, and flexibility.

In this chapter, different techniques for hardware realization of ANNs are first discussed. These techniques include analog, digital, and mixed-mode VLSI circuits for implementing ANNs.

A mixed-mode sampled-data (SD) implementation of ANNs is then presented. A chip containing novel programmable switched-resistor (SR) synapses as well as a simple CMOS analog neuron is designed, fabricated, and tested. ANN model simulations are performed for a prototype multi-layer perceptron (MLP) model architecture which solves two character recognition problems.

The CMOS ANN circuit which implements the prototype MLP is tested using HSPICE simulations. An extended MLP model architecture is proposed to solve multi character recognition problems. A parallelogram VLSI architecture is developed to implement the CMOS ANN circuit of the prototype MLP. A novel modular ANN chip which implements a two character recognizer is designed using a 1.2 μ CMOS technology. An architecture for multi character recognition system is finally proposed.

This research demonstrates the feasibility of an SR CMOS VLSI implementation of ANNs for character recognition using the parallelogram developed VLSI architecture.

An Overview 23

1.8.2 Chapter 3: A Design Automation Environment for Mixed Analog/Digital ANNs

With the increasing popularity of Artificial Neural Network (ANN) systems for their ability to solve many engineering problems, there has been a growth in activity during the past five years in VLSI implementation of these systems. Many designs have been reported in this period, using both digital and analog circuit techniques. In order to keep up with the progress in hardware design, it is necessary to develop automation techniques for VLSI implementation of ANN systems, in order to achieve faster turn-around times and making better use of hardware resources. In this respect, a Design Automation Environment is proposed in this chapter, with the objective of helping hardware designers to quickly, but systematically, investigate the various implementation choices that are available, by using a combination of automatic analysis and synthesis methods. The environment is targeted for generalized descriptions of popular ANN systems with diverse learning procedures, connection structures, and performance requirements. The ANN systems are input to the environment with the help of Data Flow Graph (DFG) descriptions. The output of the environment is a high-level interconnection description of analog or mixed analog/digital circuit blocks. The components of the environment include two execution frameworks called The Analyzer and The Synthesizer, respectively, and a Design Library. The effect of circuit non-idealities on the performance of the ANN systems is quantitatively evaluated by The Analyzer. A combination of techniques that use Taylor series approximation and non-linear optimization, are applied automatically to the DFG description, and a set of specifications are generated, that set limits on permissible amounts of error due to non-ideal effects from various operations. The Design Library stores models of non-ideal behavior from various building block circuits. The Synthesizer uses the limits on errors generated by The Analyzer, and consults these behavioral models for screening the circuits that are suitable for implementing the given ANN system. If digital circuits are required for implementing certain operations, because the specifications are too strict to be met by analog circuits, the Synthesizer partitions the DFG description to reduce the interface between modules operating in these two distinct modes. The operations in the DFG are scheduled and specific hardware modules are allocated to each operation, by a new scheduling and allocation technique. This technique not only takes care of the special requirements of mixed analog/digital hardware, but also handles the regularity and parallelism of ANN systems, by improving an area-time cost function by selectively sequentializing various operations. Special heuristics are added to the Synthesizer to embed hardware that are intended to be operated asynchronously. The utilities of the Design Automation

Environment are demonstrated with the help of design examples from three different ANN systems, namely (a) the Multi-Layer Perceptrons (MLP) that uses the Back-Propagation algorithm for learning (b) the Hopfield Networks and (c) multi-layer ANNs that uses the Mean-Field Theory for learning.

1.8.3 Chapter 4: A Compact VLSI Implementation of Neural Networks

This chapter reports on a class of compact circuit building blocks that has been designed to implement the neural networks with digital inputs and outputs and analog weights. A modified feedforward neural network with learning (FNN) has been implemented by the designed building block circuits. This network which is applicable to encoder problems has been tested in the functional and circuit levels. A VLSI design system prototype MANNA for FNN implementations has been developed based on a two-step mapping method. Given the neural network specifications, MANNA automatically generates FNN chips. First, the FNN architectures are mapped into FNN circuits, then further mapped into FNN chips. At the microstructure level, analog cell design methodologies effectively implement low precision neural computation. In the macrostructure level, digital design automation methodologies are adapted to implement FNNs which posses a very high degree of modularity.

1.8.4 Chapter 5: An All-Digital VLSI ANN

For artificial neural networks (ANNs) to take full advantage of state-of-the-art VLSI and ULSI technologies, they must adapt to an efficient all-digital implementation. This is because these technologies, at higher levels of integration and complexity, are mainly a digital implementation medium, offering many advantages over analog counterparts. This chapter illustrates this theory by adapting one of the most complicated ANN models, the neocognitron (NC), to an efficient all-digital implementation for VLSI. The new model, the Digi-NeoCognitron (DNC), has the same pattern recognition performance as the NC. The DNC model is derived from the NC model by a combination of preprocessing approximations and the definition of new model functions, e.g., multiplication and division are eliminated by conversion of factors to powers of 2, requiring only shift operations. In this chapter, the NC model is reviewed, the DNC model is presented, a methodology to convert NC models to DNC models is discussed, and the performance of the two models are compared on

An Overview 25

a character recognition example. The DNC model has substantial advantages over the NC model for VLSI implementation. The area-delay product is improved by two to three orders of magnitude, and I/O and memory requirements are reduced by representation of weights with 3 bits or less and neuron outputs with 4 bits or 7 bits.

1.8.5 Chapter 6: Neural Predictive Hidden Markov Models for Automatic Speech and Speaker Recognition

Speech is certainly the most natural and efficient form of human communications. For the past several decades researchers in the field of automatic speech recognition (ASR) have been driven by the goal of establishing a means through which people can talk with computers in much the same way they carry on conversations with fellow humans. Some breakthroughs have been achieved in this area utilizing statistical based classifiers. However, this success appears really limited if the abilities of current ASR machines are compared to those of a five year old child.

Recently a lot of research has focused on the application of artificial neural networks in the area of speech recognition. Interest in this computing paradigm has been fueled by several factors. These include their strong discriminative abilities allowing for high accuracy speech recognition. Also, their massive parallelism makes them well suited for parallel VLSI implementation. In addition, they mimic the human mode of learning by example and generalizing thereafter, to novel inputs.

This chapter starts by presenting an overview of the Automatic Speech Recognition (ASR) problem and the various techniques utilized for tackling it. Next a hybrid ASR system utilizing both MLPs as pattern predictors and Hidden Markov Models as time warping tools is presented. An in depth analysis is then given illustrating how rigorous theoretical analysis could be a definite advantage in arriving at optimal structures of neural networks to accurately model the speech signal. Comparative speech recognition experiments are also reported to indicate the superiority of the developed system to other models.

A corrective scheme for training the above model is then outlined and compared to standard non-corrective training schemes from a speech accuracy point of view. Finally the feasibility of utilizing this model as an accurate text dependent speaker recognition system is demonstrated.

1.8.6 Chapter 7: Minimum Complexity Neural Networks for Classification

In this chapter a framework for designing optimal probabilistic-based neural networks for probability density estimation and classification is presented. The framework is based on two pillars. Firstly, a Bayesian optimal model selection approach is employed for selecting minimum complexity models. Secondly, optimal learning criteria are developed for training the models. In the context of Probability Density Function (PDF) estimation, a Discrete Stochastic Complexity criterion "DSC" is developed from the Bayesian framework, which is used in conjunction with optimal Maximum Likelihood "ML" estimation. The DSC/ML technique is applied to both Gaussian mixture models and hard competitive models for PDF estimation and clustering. The proposed technique is capable of determining the number and shape of clusters in the data, making it very useful in vector quantization codebook designs. In the context of supervised classification, three novel neural network classifiers are proposed. These are namely, the Adaptive Probabilistic Neural Network "APNN", the Adaptive Nearest Neighbor Classifier "ANNC", and the Adaptive Feature extraction Nearest Neighbor classifier "AFNN". For each classifier, a Discrete Stochastic Complexity Criterion "DSCC" for classification is developed from the optimal Bayesian framework. Optimal discriminative criteria are developed for these classifiers, namely the Maximum Mutual Information "MMI" and the Least-Mean-Squared Error "LMSE", and their learning algorithms are derived. The proposed classifiers are tested on many pattern recognition problems, including a vowel recognition task with reported results. The proposed classifiers offer substantial reduction of storage and computational complexity over other compared classifiers, as well as excellent recognition results.

1.8.7 Chapter 8: A Fuzzy Clustering Neural Network (FCNN) for Pattern Recognition

This chapter presents a fuzzy clustering neural network (FCNN) model which uses Gaussian nonlinearity. A learning algorithm, based on direct fuzzy competition between the nodes, is presented. The connecting weights, which are adaptively updated in batch mode, converge towards values that are representative of the clustering structure of the input pattern. Mapping the proposed algorithm onto the corresponding structures in VLSI medium, the FCNN architecture with three types of processing cells is developed so that on-line learning and parallel implementation are feasible. The effectiveness of the FCNN is illustrated by applying it to a number of test data sets and the result is compared to the performance of the FCM algorithm.

1.8.8 Chapter 9: A Pipelined Architecture for Neural-Network-Based Speech Recognition

Artificial neural networks (ANNs), as processors of time-sequence patterns, have been successfully applied to several speaker-dependent speech recognition systems. This chapter develops efficient ANN pipelined architectures, which include both parallel and serial data flow processing stages. The implementations of these stages are efficiently matched to the VLSI medium. Compared with the typical ANN structure models for speech recognition, e.g., time-delay neural networks (TDNN), block-windowed neural network (BWNN) and dynamic programming neural network (DNN), the proposed architecture has the lowest hardware complexity while maintaining a high throughput rate. The performance of the architecture is analyzed and its effectiveness is illustrated.

REFERENCES

[1] S.Y. Kung; "Digital Neural Networks", PTR Prentice Hall, 1993.

[2] J.D. Cowan and D.H. Sharp; "Neural Nets", Technical Report, Mathematics Department, University of Chicago, 1987, pp. 2.

[3] F. Rosenblatt; "The Perceptron: A Probabilistic Model for Information Storage and Organization in the Brain", Psychological Review, 65, pp. 386-408, 1958.

[4] M.D. Richard and R.P. Lippmann; "Neural Network Classifiers Estimate Bayesian a Posteriori Probabilities", Neural Computation, 3(4), pp. 461-483, 1991.

[5] B. Widrow and M.E. Hoff; "Adaptive Switching Circuits", In 1960 IRE WESCON Convention Record, pp. 96-104, New York, NY, 1960.

[6] R.O. Duda and P.E. Hart; "Pattern Classification and Scene Analysis". Wiley, New York, NY, 1973.

[7] R.P. Lippmann; "An Introduction to Computing with Neural Nets", IEEE Acoustics, Speech and Signal Processing Magazine, 4(2), pp. 4-22, April 1987.

[8] J. Makhoul, A. El-Jaroudi, and R. Schwartz; "Formation of Disconnected Decision Regions with a Single Hidden Layer". In Proceedings of the International Joint Conference on Neural Networks, Vol.1, pp. 455-460, 1989.

[9] D.L. Chester; "Why Two Hidden Layers are Better than One". In Proceedings of the International Conference on Neural Networks, Vol.1, pp. 265-268, Erlbaum, 1990.

[10] A. Hajnal, W. Maass, P. Pudlak, M. Szegedy, and G. Turan; "Threshold Circuits of Bounded Depth". In Proceedings of the 1987 IEEE Symposium on the Foundations of Computer Science, pp. 99-110, 1987.

[11] W. Fakhr and M.I. Elmasry; "A Fast Learning Technique for the Multilayer Perceptron". IEEE International Joint Conference on Neural Networks, San Diego, 1990.

[12] D.B. Parker; "Learning Logic". Technical Report TR-47, Center for Comp. Res. in Econ. and Man., MIT, Cambridge, MA, April 1985.

[13] D.E. Rumelhart, G.E. Hinton, and R.J. Williams; "Learning Internal Representations by Error Propagation". In D.E. Rumelhart and J.L. McClelland, editors, Parallel Distributed Processing: Explorations in the Microstructure of Cognition: Foundations, MIT Press, Cambridge, MA, 1986.

[14] P.J. Werbos; "Beyond Regression: New Tools for Prediction and Analysis in the Behavioral Science". Doctoral Dissertation, Applied Mathematics, Harvard University, Boston, MA, November 1974.

[15] G. Cybenko; "Approximation by Superpositions of a Sigmoidal Function", Mathematics of Control, Signals, and Systems, 2(4), pp. 303-314, 1989.

[16] S.E. Fahlman and C. Lebiere; "The Cascade-Correlation Learning Architecture", In D. Touretzky, editor, Advances in Neural Information Processing Systems 2, pp. 524-532. Morgan Kaufmann, 1990.

[17] A.R. Barron and R. Barron; "Statistical Learning Networks: A Unifying View", In E.J. Wegman, D.I. Gantz, and J.J. Miller, editors, Computing Science and Statistics: Proc. of the 20th Symposium on the Interface, pp. 192-202, 1989.

[18] A.C. Ivakhnenko; "Polynomial Theory of Complex Systems", IEEE Transactions on Systems, Man, and Cybernetics, 1, pp. 364-378, 1971.

[19] J.H. Friedman and W. Stuetzle; "Projection Pursuit Regression", J. Amer. Stat. Assoc., 76, pp. 817-823, 1981.

[20] J.H. Friedman and J.W. Tukey; "A Projection Pursuit Algorithm for Exploratory Data Analysis", IEEE Transactions on Computers, 23, pp. 881-889, 1974.

[21] Y. le Cun, J.S. Denker, and S.A. Solla; "Optimal Brain Damage". In D. Touretzky, editor, Advances in Neural Information Processing Systems 2, pp. 598-605. Morgan Kaufmann, 1990.

[22] B. Hassibi and D.G. Stork; "Second Order Derivatives for Network Pruning: Optimal Brain Surgeon". In Advances in Neural Information Processing Systems 5. Morgan Kaufmann.

[23] M.J.D. Powel; "Radial Basis Functions for Multivariate Interpolation: A Review". Technical Report DAMPT 1985/NA 12, Dept. of App. Math. and Theor. Physics, Cambridge University, Cambridge, England, 1985.

[24] J. Moody and C.J. Darken; "Learning with Localized Receptive Fields". In Proceedings of the 1988 Connectionist Models Summer School, pp. 133-143, 1988.

[25] J. Moody and C.J. Darken; "Fast Learning in Networks of Locally-Tuned Processing Units", Neural Computation 1, pp. 281-293, 1989.

[26] T. Poggio and F. Girosi; "A Theory of Networks for Approximating and Learning". Artificial Intelligence Lab. Memo 1140, MIT, 1989.

[27] J.T. Tou and R.C. Gonzalez; "Pattern Recognition Principles". Addison-Wesley, Reading, MA, 1974.

[28] K. Fukunaga; "Introduction to Statistical Pattern Recognition". Academic Press, San Diego, CA, 1972.

[29] J.A. Hartigan; "Clustering Algorithms". Wiley, New York, 1975.

[30] T. Kohonen; "An Introduction to Neural Computing". Neural Networks, 1(1), pp. 3-16, 1988.

[31] T. Kohonen; "The Self-Organizing Map", Proc. IEEE, Vol.78, pp. 1464-1479, 1990.

[32] L. Breiman, J.H. Friedman, R.A. Olshen, and C.J. Stone; "Classification and Regression Trees", Wadsworth and Brooks, Pacific Grove, CA, 1984.

[33] J.S. Albus; "Mechanisms of Planning and Problem Solving in the Brain". Mathematical Bioscience, 45, pp. 247-293, 1979.

[34] J. Hertz, A. Krogh, and R.G. Palmer; "Introduction to the Theory of Neural Computation". Addison-Wesley, Redwood City, CA, 1991.

[35] T.J. Sejnowski and C.R. Rosenberg; "NETtalk: A Parallel Network that Learns to Read Aloud". Technical Report JHU/EECS-86/01, Johns Hopkins University, Baltimore, MD, 1986.

[36] K.J. Lang, A.H. Waibel, and G.E. Hinton; "A Time-Delay Neural Network Architecture for Isolated Word Recognition", Neural Networks, 3(1), pp. 23-44, 1990.

[37] J. Robinson and F. Fallside; "Static and Dynamic Error Propagation Networks with Application to Speech Coding". In Proceedings, IEEE Conference on Neural Information Processing Systems, pp. 632-641, Denver, 1988.

[38] R.L. Watrous and L. Shastri; "Learning Phonetic Features using Connectionist Networks: An Experiment in Speech Recognition". In Proceedings of the IEEE International Conference on Neural Networks, pp. 381-388, San Diego, CA, June 1987.

[39] T. Kohonen; "Self-Organization and Associative Memory", Series in Information Science, Vol.8, Springer-Verlag, New York, 1984. [40] G.A. Carpenter and S. Grossberg; "ART2: Self-Organization of Stable Category Recognition Codes for Analog Input Patterns". In Proceedings, IEEE International Conference on Neural Networks, pp. 727-736, San Diego, 1987.

[41] K. Fukushima; "A Neural Network for Visual Pattern Recognition". IEEE Computer Magazine, 21(3), pp. 65-76, March 1988.

[42] W. Fakhr, M. Kamel and M.I. Elmasry; "Unsupervised Learning by Stochastic Complexity". WCNN'93, Portland, 1993.

[43] S.Y. Kung; "Adaptive Principal Component Analysis via an Orthogonal Learning Network". In Proceedings of the International Symposium on Circuits and Systems, pp. 719-722, New Orleans, May 1990.

[44] S.Y. Kung and K.I. Diamantaras; "A Neural Network Learning Algorithm for Adaptive Principal Component Extraction (APEX)". In Proceedings of the IEEE International Conference on Acoustics, Speech, and Signal Processing, pp. 861-864, Albuquerque, NM, April 1990.

[45] L.I. Burke; "Clustering Characterization of Adaptive Resonance". Neural Networks, Vol.4, pp. 485-491, 1991.

[46] K. Fukushima, S. Miyaki, and T. Ito; "Neocognitron: A Neural Network Model for a Mechanism for Visual Pattern Recognition". IEEE Transactions on Systems, Man and Cybernetics, SMC-13. pp. 826-834, 1983.

[47] J.J. Hopfield and D.W. Tank; "Neural Computation of Decision in Optimization Problems". Biological Cybernetics, 52, pp. 141-152, 1985.

2

A SAMPLED-DATA CMOS VLSI IMPLEMENTATION OF A MULTI-CHARACTER ANN RECOGNITION SYSTEM

Sameh E. Rehan and Mohamed I. Elmasry

2.1 INTRODUCTION

Artificial Neural Networks (ANNs) are successfully applied to a variety of data classification and recognition problems. Through experimentation and simulation, acceptable solutions to such problems can be obtained using ANNs. However, the effectiveness of an ANN algorithm strongly depends on the hardware that executes it. This hardware has to capture the inherent parallelism of the ANNs and tolerate the ANN's need of massive numbers of computations.

In this chapter, a Sampled-Data (SD) VLSI ANN two-character recognition system is developed using novel building block circuits and an optimal VLSI parallel architecture. This optimal VLSI architecture is modular for easy expansion to larger networks as well as flexible to allow single chips to be joined together to form very large networks.

A SD VLSI ANN multi-chip multi-character recognition system is also proposed. The procedure used to implement this ANN multi-character recognition system is summarized in the following steps:

- Novel programmable switched-resistor (SR) synapse and simple analog neuron circuits are developed, fabricated, and tested.
- The characteristics of the developed synapses and neurons are encoded in the Back-Propagation (BP) learning algorithm of the multi-layer perceptron (MLP) model simulator.
- A two-character MLP ANN model architecture is selected as a prototype to test these implementation steps.

- Using the prototype MLP model architecture, ANN model simulations are performed to obtain a set of weights that can be used to distinguish between the two characters T and C.
- HSPICE circuit simulations are conducted to verify the validity of the ANN circuit to perform the desired recognition task.
- An optimal VLSI architecture to implement the prototype MLP is then developed.
- The layout of a two-character ANN recognizer is designed using a 1.2 μm CMOS technology and HSPICE simulations of the extracted circuit are performed to check the functionality of the designed chip.
- The fabricated chip is tested and the experimental results are compared with the HSPICE simulation results.
- A modular two-character MLP ANN model architecture is introduced and used as a basic building block for a multi-character ANN recognition system.
- A multi-chip multi-character ANN recognition system is proposed using multiples of the modular version of the developed two-character recognizer and an output-stage ANN chip.
- The layout of both the modular two-character recognizer and the output-stage chips are designed using a 1.2 μm CMOS technology and HSPICE simulations of the extracted circuits are performed to check the functionality of the designed chips.
- The fabricated chips are tested and the experimental results are compared with the HSPICE simulation results.

In the next section, different ANN implementation techniques are briefly described. In the following sections, the above implementation steps of a SD VLSI ANN multi-character recognition system are explained in detail.

2.2 ANN IMPLEMENTATION TECHNIQUES

2.2.1 Historical Perspective

The first physical ANN implementation was created by Minsky and Edmonds in 1951 [1]. They designed a 40-neuron learning machine based upon the Hebbian learning rule. The machine successfully modeled the behavior of a rat in a maze searching for food. Hay et al. built MARK I, a 512-neuron pattern recognition system, that learned using the basic perceptron learning rule in 1960 [2]. During the same year, Widrow introduced an implementation for his adaline [3] and Crane introduced a neurally inspired transistor called the neuristor [4].

During the period from the 1960's to the mid 1980's, significant progress has been achieved in theoretical aspects of ANNs. However, ANN concepts and applications are usually tested through simulations of neural algorithms on digital computers. This trend represents the most popular ANN implementation medium till today because of the computer's versatility, cost, and ever-increasing capability. In addition, the advent of massively parallel computers, supercomputers, and specialized accelerator boards has provided even greater capabilities in these areas.

In 1984, Hecht-Nielsen and Gutschow developed the TRW MARK III, a neurocomputer that could be used to implement a wide variety of ANN models [5]. At the same time, Farhat et al. introduced an optical implementation of the Hopfield mode [6]. Since 1986, more ANN electronic implementations were developed by several major research laboratories (e.g. AT&T, CalTech, MIT/Lincoln Lab.) [7].

The broad range of ANN implementations that exist today can be placed into a four class taxonomy, shown in Figure 2.1, that includes:

- Analog Implementations – defined as the ANN implementations which use only analog computations and storage.

- Optical Implementations – defined as the ANN implementations that involves the use of optical components.

- Digital Implementations – defined as the ANN implementations using digital computers/integrated circuits.

- Mixed-mode Analog/Digital Implementations – defined as the ANN implementations utilizing both analog and digital techniques.

In the following sub-sections, each of these classes and their subclasses are described in more detail.

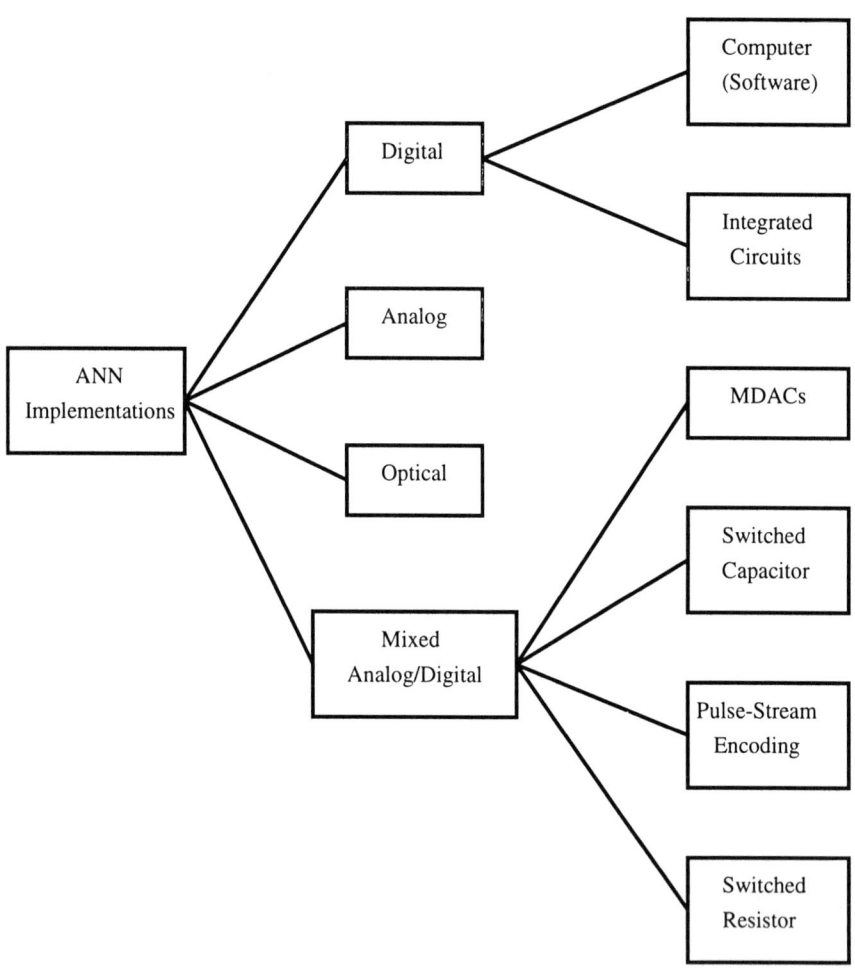

Figure 2.1 Taxonomy of ANN implementations

2.2.2 Analog Implementations

Most of the microelectronic implementations use some form of analog computations [8–14].

In analog VLSI ANNs, the synapse weight information can be stored in the dynamically refreshed capacitors for medium-term storage or in the floating-gate of an EEPROM cell for long term storage.

Advantages:

The main benefit of the analog approach is that a wide range of operations can be performed using a small number of transistors. This means simple basic blocks and connection, which leads to small area and thus larger networks can be built on a single chip.

Drawbacks:

The major drawback of conventional analog circuits is that they have not been fully integrated in MOS technology. A major reason for that is the necessity of accurately defining resistance-capacitance products. Accurate time constants require that the absolute value of resistors and capacitors be well controlled. This condition is not found in present integrated circuit technology without trimming. For polysilicon or diffused resistors, fabrication process as well as temperature variations contribute to an uncertainty of resistor values which can be 10% or more. In addition, resistive connections are impractical to implement in VLSI form because the resistive element tends to be nonlinear and occupy too much area [15].

However, analog VLSI has been identified as a major technology for future information processing. This is primarily because some of the traditional analog design limitations such as accurate absolute component values, device matching, and precise time constants are not major concerns in ANNs. This is due to the fact that computation precision of individual neurons does not seem to be of paramount importance.

A major drawback, which is very critical in ANNs, is that conventional analog implementations are fixed (i.e., no programmability can be achieved). Another main drawback is the lack of a reliable and nonvolatile memory. A solution to this problem is to use floating-gate technologies [16]. This requires high voltages on the chip, and the aging problems have yet to be resolved.

2.2.3 Optical Implementations

Using electrical conducting wires, in a two dimensional surface, to implement the ANN interconnections has two main limitations. The first one is that these interconnections can not cross each other and the second is that a minimum space should be left between these conducting paths to prevent the interference between the signals in adjacent interconnections. This leads to a limited number of interconnections using a two dimensional electrical implementation.

Optical ANN computers have two main components. The first component is the optical switching element which simulates the neuron. The optical switching elements, arranged in two dimensional arrays, can be purely optical or a combination of optics and electronics (using a light detector, an electronic switch, and an LED [5]), where nonlinear elements can be fabricated using gallium arsenide. The second element of the optical ANN computers is the hologram which implements the interconnection matrix. A one cubic centimeter of a hologram can have more than one trillion connections [17].

Although optical ANNs are promising for the future, they are difficult to implement with today's technology, especially the nonlinear optical switches. In addition, the current resolution of the transparencies used for the weight interconnection matrix is quite low [18]. Also, almost all optical systems are bulky and experimental [19].

2.2.4 Digital Implementations

Conventional digital techniques use software simulation of the network dynamics on general purpose computers, parallel computers, or workstations provided with hardware accelerators. However, any general-purpose digital computer is an order of magnitude slower than neural hardware, which could be directly produced using the same fabrication technology as the digital computer [20].

2.2.4.1 Software implementations

Conventional computers as simulators:

ANNs were originally studied by simulating its algorithms in a traditional computing environment. In theory, conventional computers can implement any size of ANN with any number of interconnections. Unfortunately; their performance, when running ANN algorithms, is very low ($5 * 10^7$ IC/S for Cray X-MP supercomputer).

Even these low performance figures obtained from benchmark problems are optimistic and misleading, since they cannot be achieved for medium or large ANNs. This is because of the storage hierarchy of conventional computers. They keep the most important data in the high speed cache memory, while the rest of the data is kept in a slower memory. Benchmark simulations often use small networks which can execute in the cache. This is not possible for large ANNs and, as a result, the performance degrades.

Parallel computers as simulators:

Because the nervous system has numerous neurons working in parallel, and parallel computers have multiple processors working in parallel; it has been reasoned that parallel computers are the best way to run ANN paradigms. Unfortunately, it was discovered that parallel computers are not suitable for ANN simulations. This is because they are good at increasing the amount of computation performed at processor nodes but the majority of calculations in ANNs are performed in the communication channels. For example, a fully connected ANN with 100 neurons requires 100 node calculations and 10,000 communication channel calculations (multiplication and forwarding).

With massively parallel processing capabilities, ANNs can be used to solve many engineering and scientific problems. However, their ANN implementations are programmed in languages that are often unique to each machine and offer very little portability. The processing speed of these implementations is slightly-slower to much-slower than supercomputer implementations, but they have the valuable ability to interact with the ANN as it is processing.

Coprocessors/Accelerator boards:

Many digital neural coprocessors, which usually interface to personal computers and engineering workstations, are commercialized for accelerating neurocomputation. Typical products include ANZA from Hecht-Nielson Co., SAIC from Sigma, Odyssey from Texas Instruments Inc., and Mark III/IV from TRW Inc. [19].

Dedicated ANN Accelerator boards have the best cost performance for general ANN systems. They use the host computer for the data input/output, user interface, and data storage. They are implemented using off-the-shelf digital technology. They usually have a general purpose CPU for housekeeping and a multiplier unit. The multiplier unit is either a high performance multiplier or a DSP chip.

These digital accelerators, used to simulate ANNs, are now commercially available; and provide magnitudes of improvement in the processing speed of ANN applications. However, they are still orders of magnitude slower than what we can achieve by directly fabricating a network with hardware [8].

In conclusion; software simulations constitute a bottleneck in practical applications, where training time can take days or weeks. Software simulations of ANNs are much slower in comparison with signal processing by circuits (hardware). Therefore, the practical use of ANNs is heavily based on circuit implementations. The most promising approach for implementing ANNs is to fabricate special-purpose VLSI chips.

2.2.4.2 Digital integrated circuits

Numerous digital chip architectures are suggested for ANNs [21–26].

Advantages:

A digital VLSI chip can be designed with available CAD tools; standard cells and libraries can be used, and digital chips are tolerant to parameter dispersion and noise. Extremely high speed (bit rates in excess of 100MHz) and arbitrarily high precision can be achieved. Its storage devices are simple and highly reliable.

Drawbacks:

An all digital network would require a complex processor for each coupling element, resulting in an impractically large circuit even for a modest number of neurons on a chip [9]. Also, a large amount of memory is necessary to store the weights of synapses as well as neuron values with an additional area penalty. Moreover, one of the unique features of neural networks (they are fault-tolerant) is lost to some extent because digital implementations consume a lot of area for an accuracy which is not always needed.

Also, a general-purpose digital neural coprocessor is usually much slower than a special-purpose analog neural hardware which implements the neural network in an optimal fashion.

2.2.5 Mixed Analog/Digital Implementations

Fully digital implementation of ANNs do not capture the unique features of biological NNs, such as inherent parallelism, compactness, and simplicity. On the other hand, fully analog implementation lacks reliable permanent storage and controllability required for multiplexing. So, we propose using a mixed-mode implementation where the advantages of both these techniques can be combined. However, the analog circuits must be compatible with what is essentially digital integrated circuit process technology.

The basic motivation for the development of the SD techniques was the need to obtain fully-integrated high-quality devices which require no trimming and use as little silicon circuit area as possible. One of the primary advantages of SD circuits is that they provide a means of economically and accurately implementing analog circuit functions and provide the possibility of continuously tuning the characteristics of a circuit by digital means which is needed in many circuit applications.

2.2.5.1 Multiplying digital-to-analog converters (MDACs)

One approach, proposed and investigated in [27] and [28], is to use multiplying digital-to-analog converters (MDACs). In this approach, the weights are stored digitally in RAM cells. The weights are transferred directly from the RAM cells to the MDACs in their digital form and the signals are kept in their analog form. Thus, the multiplication between the analog signal and the digital weight can be done directly without using A/D or D/A converters. However, this approach still offers a high VLSI design cost for medium and large ANNs. In [27], 1024 4-bit MDACs are implemented using 28,500 transistors. The active area occupies $28mm^2$ using 3-micron design rules.

2.2.5.2 Switched-Capacitor (SC) techniques

Recently, VLSI implementations of ANNs using SC techniques were proposed. The proposed techniques are adequate for solving optimization problems [29], linear and nonlinear programming problems [15] and [30], and speech recognition [31]. SC summing integrator, with a binary weighted capacitor array at its input, is proposed to implement an ANN with limited connectivity and limited precision [32]. This proposed SC ANN circuit can achieve a useful classification rate, for pattern recognition problems, which exceeds that of powerful conventional classifiers at potentially very high speed. A time-multiplexed SC circuit is proposed in [33] for ANN applications.

Advantages:

Since the insulator in a properly fabricated MOS capacitor has essentially ideal characteristics, much better stability and linearity are obtained for SC resistors than is possible with diffused resistors. In addition, the ratio of two MOS capacitors has very little temperature dependence and can be implemented with the accuracy of 0.1 percent. Also, SC techniques can directly implement the ANN with less area than the resistive connections.

SC circuits have the ability to invert the signal without requiring an additional inverter. In contrast to other electronic neuron circuits, which have a single type of input and use two inverters to provide inverting and noninverting outputs, the SC implementation of an artificial neuron has a single output and inputs which can be inverted by using a suitable SC element.

An important advantage of SC circuits is that they can be digitally controlled to execute the programmable function. SC circuits exhibit a very high modularity as well as simple analog processors. SC circuit structures provide the immunity to the variation of process and temperature. So, SC circuits have the advantage of VLSI implementation and programmability [15].

Drawbacks:

While the SC resistors are exactly equivalent to resistors by themselves, such an equivalence may not hold true when the realizations of SC elements are used to replace resistors of a network. Also, SC circuits are limited by essentially the time required to completely transfer the charge from one capacitor to another.

SC circuits use large selectable-capacitor arrays to achieve programmability. In the most general case; many such arrays are required, consuming a lot of chip area. Even if the capacitor arrays were to be time-shared, they would have to be large in order to obtain sufficient resolution; furthermore, such resolution is fixed once the circuit has been fabricated and there is no degree of freedom associated with it. A major drawback of existing SC techniques is that under VLSI scaling, capacitor ratio accuracy may deteriorate in certain fabrication processes [34].

Another factor which must be considered for some SC circuits is that they require a sample-and-hold circuit at the input to avoid continuous signal feed-through that causes frequency response distortion.

2.2.5.3 Pulse-Stream encoding technique

In this technique, a pulse stream signaling mechanism is used by the neural circuitry. The process is analogous to that found in natural neural systems; where a neuron that is ON fires a regular train of voltage spikes on its output, while an OFF neuron does not. This technique uses a hybrid approach which tries to blend the merits of both digital and analog technologies by performing analog multiplication under digital control [35] and [36].

Advantages:

Analog multiplication is attractive in neural VLSI, for reasons of compactness, potential speed, and lack of quantization effects. On the other hand, avoiding digital multiplication is preferred because it is area- and power-hungry.

Digital signals are robust, easily transmitted and regenerated, and fast. On the other hand, analog signals are far from robust against noise and interference, and are susceptible to process variations between devices.

Drawbacks:

The summation in the general ANN equation $\{\sum_{i=1}^{N} w_{ij} \cdot a_i\}$ is not the result of N simultaneous multiplications and additions. In this technique; the summation is performed sequentially, i.e. it is distributed in time.

The synaptic weights are restricted to be from -1 to 1. This restriction limits the possibility to reach a convergence in learning. Also, each synapse has to have two input lines, one for the excitatory inputs and the other for the inhibitory inputs.

2.2.5.4 Switched-Resistor (SR) technique

Digital tuning can be accomplished by using MDACs or by changing capacitor ratios in SC circuits. These techniques are expensive and require a great deal of integrated circuit area. For this reason, several authors have suggested periodically switching resistors as a simple means of tuning circuits [37–38].

In [39], Ming-Lei Liou has suggested that a MOS transistor may be used for the realization of the switch, together with the resistor, which may be considered as the switch-on resistance. This realization can be achieved by applying a clock signal to the gate of the MOS transistor. The resistance of the MOS varies by varying the duty-cycle of the clock. A simple proof for this is given in [40]. Referring to Figure 2.2, the average resistance of the MOS transistor is given by

$$R_{av} = \frac{V_S - V_D}{I_{av}} \qquad (2.1)$$

(a)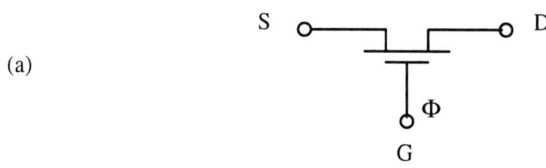

(b)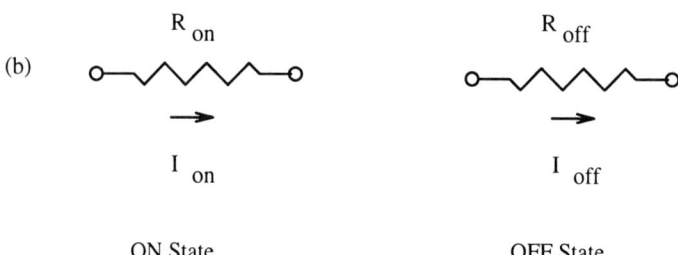

(c)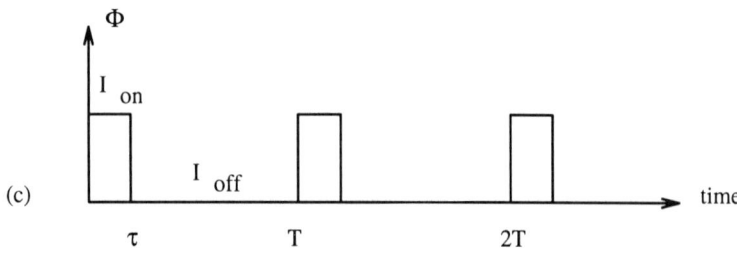

Figure 2.2 (a) An NMOS transistor (b) Equivalent resistance (c) Controlling clock

where, I_{av} is the average current passing through the MOS during the whole period of time T. I_{av} can be calculated as follows

$$I_{av} = \frac{1}{T}\left[\int_0^\tau I_{on}\, dt + \int_\tau^T I_{off}\, dt\right] \quad (2.2)$$

$$I_{av} = \frac{1}{T}[I_{on}\cdot\tau + I_{off}\cdot(T-\tau)]$$

$$I_{av} = \frac{1}{T}\left[\frac{V_S - V_D}{R_{on}}\cdot\tau + \frac{V_S - V_D}{R_{off}}\cdot(T-\tau)\right]$$

$$I_{av} = \frac{1}{T}\left[\frac{V_S - V_D}{R_{on}}\right]\cdot\left[\tau + \frac{R_{on}}{R_{off}}\cdot(T-\tau)\right] \quad (2.3)$$

As $R_{off} \Rightarrow \infty$, then equation (2.3) becomes

$$I_{av} = \frac{\tau}{T}\left[\frac{V_S - V_D}{R_{on}}\right] \quad (2.4)$$

From equations (2.1) and (2.4), we have

$$R_{av} = R_{on}\cdot\frac{T}{\tau} \quad (2.5)$$

$$G_{av} = G_{on}\cdot\frac{\tau}{T} \quad (2.6)$$

Therefore, SR technique operates on a pulse width control principle in which the value of each element is determined by the duty-cycle of a single digitally controlled analog transmission gate. In [34] and [41], Y. Tsividis used the same principle to build timing-controlled switched analog filters with full digital programmability.

Advantages:

SR elements are simpler in construction, less complicated in control, and have smaller areas compared with MDACs. Compared to SC elements; SR elements can replace resistors of any network without restrictions. Also, there is no need for capacitors to simulate the resistor which leads to a more compact element.

By using an accurate and stable digital clock with varying duty-cycles, the equivalent resistance of the analog switch can be controlled very accurately.

Also, we can use parallel connections of MOS transistors with different ON-resistance to achieve higher accuracy.

SR technique relies on timing rather than element ratios. It may actually improve in performance with scaling. SR technique as a timing-controlled technique exhibits superior programming flexibility.

Drawbacks:

Due to the parasitic capacitances between the substrate and each of the source and drain of the MOS transistor, there will be a clock feed-through noise; i.e. coupling clock from gate to source or drain.

2.3 DEVELOPED CMOS CIRCUITS FOR ANNS

Models of artificial neural networks (ANNs), with only two types of building blocks (neurons and synapses), are simple compared to conventional computing devices. A synapse can be considered as a multiplier of the incoming signal value by the stored weight value. A neuron adds together the output values of the connected synapses and performs a nonlinear function (e.g. tanh) for the resulting sum.

For efficient VLSI implementation of ANNs, the basic building blocks should be simple and small in area so that large networks can be implemented using a single chip. Also, programmable synapses give a much wider range of applications for the implemented chips.

2.3.1 The Programmable Synapse

The main obstacle in VLSI monolithic implementation of ANN models is the implementation of programmable synapses. It is conceivable to achieve variable synapse strengths by using an array of elements for each synapse, which can be programmed digitally [42]. MDACs [28], switched-capacitor (SC) resistors [29], and timing-controlled switches [34], which provide the possibility of adjusting the weights digitally, are proposed to overcome this obstacle. This, however, results in either a limited choice of possible synapse strengths, or a large chip area to accommodate high-resolution synapses.

To achieve high resolution with small chip area, Y. Tsividis proposed using a MOS transistor in [43]. This technique relies on the fact that the conductance of a transistor can be changed by changing the transistor's bias point. This is achieved by applying a variable voltage source, V_{GS}, on the gate of the MOS transistor. If V_{GS} is large in comparison to the drain-source voltage, V_{DS}, the nonlinearity of the synapses will be low. MOS transistors will resemble resistors of conductance G_{MOS} given by

$$G_{MOS} = \mu_n C_{ox}(W/L) \cdot (V_{GS} - V_T), \qquad (2.7)$$

where all symbols have their usual meaning.

This technique, however, suffers from the practical limitations of achieving very small voltage steps to accomplish the required resolution. In this chapter, we propose using a switched-resistor (SR) element as a programmable synapse.

2.3.1.1 The SR element

A SR synapse can be implemented using a single MOS transistor with a clock signal applied to its gate. As discussed earlier, the equivalent conductance of the MOS transistor is given by equation (2.6)

$$G_{av} = G_{on} \cdot \frac{\tau}{T}.$$

One important nonideal characteristic of the MOS transistor is the parasitic capacitances between the substrate and each of the source and the drain of the MOS transistor. Due to these parasitic capacitances, there will be a clock feed-through noise; i.e. coupling clock from gate to source or drain [44].

A better implementation of a SR synapse is to use a transmission gate (or pass transistor). Transmission gates (TGs) are constructed using one NMOS and one PMOS transistor in the parallel connected arrangement shown in Figure 2.3. The TG acts as a bidirectional switch that is controlled by the gate signal C. When $C = 1$; both MOSFETs are ON, allowing the signal to pass between points A and B. On the other hand, $C = 0$ places both MOSFETs in CUTOFF, creating a high impedance path between A and B and prohibiting all but leakage currents from flowing.

The CMOS representation of the SR synapse, TG, has two advantages over the single MOS transistor. The first is that the voltage drop across the synapse is reduced. Thus, the dynamic range is increased since smaller signals may be used.

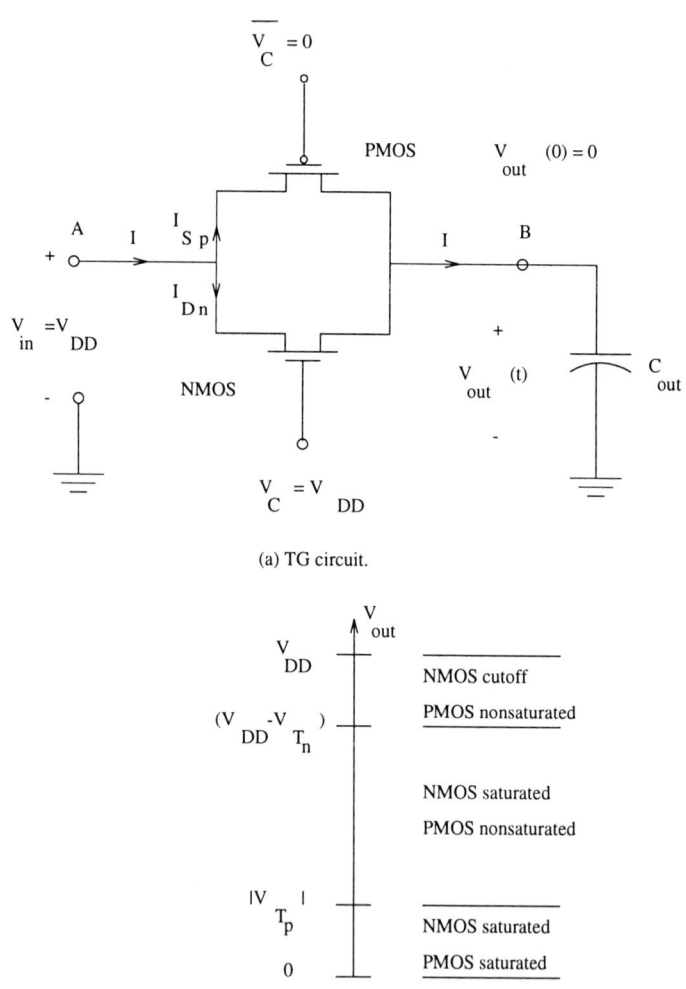

(a) TG circuit.

(b) Operation modes of the TG MOSFETs.

Figure 2.3 The TG circuit and its operation modes

The second advantage is that the clock feed-through will be diminished through cancellation. This is because the NMOS and the PMOS transistors are in parallel and require opposing clock signals.

The circuit properties of the CMOS TG, shown in Figure 2.3, can be extracted by using a simple approximation based on the concept of equivalent drain-source resistances. Equivalent resistance values depend on the MOSFET operational modes, shown in the same Figure.

To analyze the DC transmission properties of the TG, first note that the total current I through the TG is given by

$$I = I_{D_n} + I_{S_p} \tag{2.8}$$

Since the voltage across the TG is given by

$$V_{DS_n} = V_{SD_p} = (V_{DD} - V_{out}), \tag{2.9}$$

then, the total equivalent resistance at any given set of voltages can be written as follows

$$R_{eq} = \frac{(V_{DD} - V_{out})}{I}. \tag{2.10}$$

Alternately, the resistance of the individual MOSFETs can be defined by

$$R_n = \frac{(V_{DD} - V_{out})}{I_{D_n}}, \tag{2.11}$$

$$R_p = \frac{(V_{DD} - V_{out})}{I_{S_p}}. \tag{2.12}$$

Since the two are in parallel, the overall TG resistance is given by

$$R_{eq} = \frac{R_n \cdot R_p}{R_n + R_p} \tag{2.13}$$

When $V_{out} < |V_{T_p}|$ and the MOSFETs are saturated, the equivalent resistances are given by

$$R_n = \frac{2(V_{DD} - V_{out})}{\beta_n (V_{DD} - V_{out} - V_{T_n})^2}, \tag{2.14}$$

$$R_p = \frac{2(V_{DD} - V_{out})}{\beta_p (V_{DD} - |V_{T_p}|)^2}. \tag{2.15}$$

Note that $V_{T_p} = V_{TO_p}$, i.e., there is no body bias effect present, since $V_{BS_p} = 0$. The body bias of the NMOS must be included, because $V_{SB_n} = V_{out}$, using

$$V_{T_n} = V_{TO_n} + \gamma(\sqrt{2|\phi_F| + V_{out}} - \sqrt{2|\phi_F|}) \quad (2.16)$$

When V_{out} reaches $|V_{TO_p}|$, the PMOS goes into the nonsaturated region. This changes R_p to be

$$R_p = \frac{2}{\beta_p\left[2(V_{DD} - |V_{TO_p}|) - (V_{DD} - V_{out})\right]}, \quad (2.17)$$

while R_n is still given by equation 2.14.

The final transition occurs when C_{out} charges to a value $V_{out} > (V_{DD} - V_{T_n})$. The NMOS transistor will be in cutoff, which is effectively a high impedance state. The nonsaturated PMOS controls the current flow during this portion. An example of TG behavior is shown by the values plotted in the graph of Figure 2.4 [45]. It is seen that R_n increases in a manner that keeps R_{eq} relatively constant. Using the concept of equivalent resistances shows that the magnitude of I depends on the process and layout parameters. In particular, $(W/L)_n$ and $(W/L)_p$ constitute the basic design variables.

2.3.1.2 The SR programmable synapse

The schematic diagram of the developed SR programmable synapse is shown in Figure 2.5 [46]. It consists of an SR element, a switching stage, and a current inverter (CI). The circuit of the SR synapse, shown in Figure 2.6, consists of 15 elements (7 PMOS transistors and 8 NMOS transistors).

The programmable SR element operates on a pulse width control principle. A digital clock, with varying duty-cycle, is used to control the SR element's conductance which determines the magnitude of the weight. The developed SR synapse can be used to realize both positive and negative weights. The sign of the weight is implemented using the switching stage and the CI. Also, the input voltage to such a synapse can be either positive or negative. As timing-controlled circuits, SR synapses exhibit superior programming flexibility.

2.3.2 The Neuron

The schematic diagram of the developed CMOS analog neuron, which is designed to perform a near-tanh nonlinearity function, is shown in Figure 2.7. It consists of an opamp with a feedback resistor, a small capacitor, and a double inverter. The circuit of the developed neuron, shown in Figure 2.8, consists of 15 elements (5 PMOS transistors, 8 NMOS transistors, and 2 capacitors). The simple CMOS opamp is used to set a virtual ground summing point. The summed current is converted to its corresponding voltage level using the feedback resistor R of the opamp. The capacitor C is used to remove the switching effects of the controlling clocks used to determine the magnitudes of the weights connected to the neuron. The two cascaded CMOS inverters are used to perform the nonlinearity function.

2.3.3 The Controlling Unit

PLAs are used to generate the clocks required to control the values of the weights. The block diagram of a PLA designed to generate 8 clocks with different duty-cycles (3-bit resolution) is shown in Figure 2.9. Standard sequential circuit (finite state machine) design procedures may be followed to draw up a state transition table given as Table 2.1. The PLA design is given in CMOS circuit form as Figure 2.10. The same procedure can be used to achieve the required resolution.

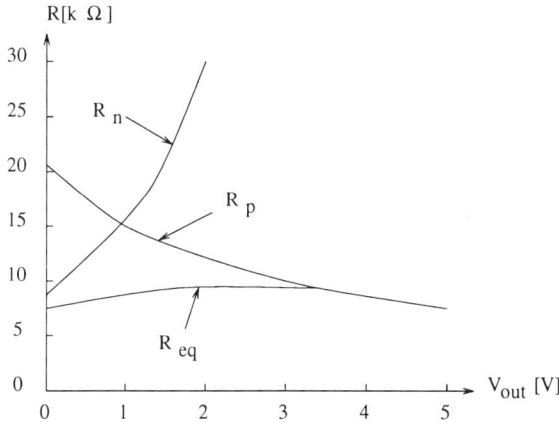

Figure 2.4 CMOS TG equivalent resistances

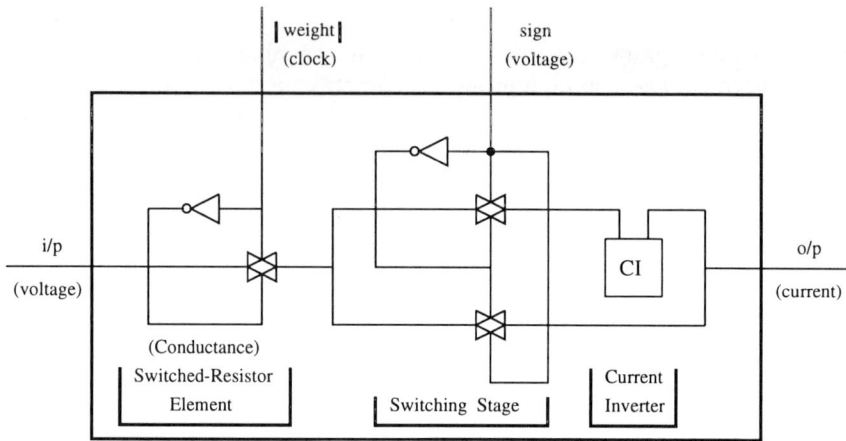

Figure 2.5 Schematic diagram of the developed SR synapse

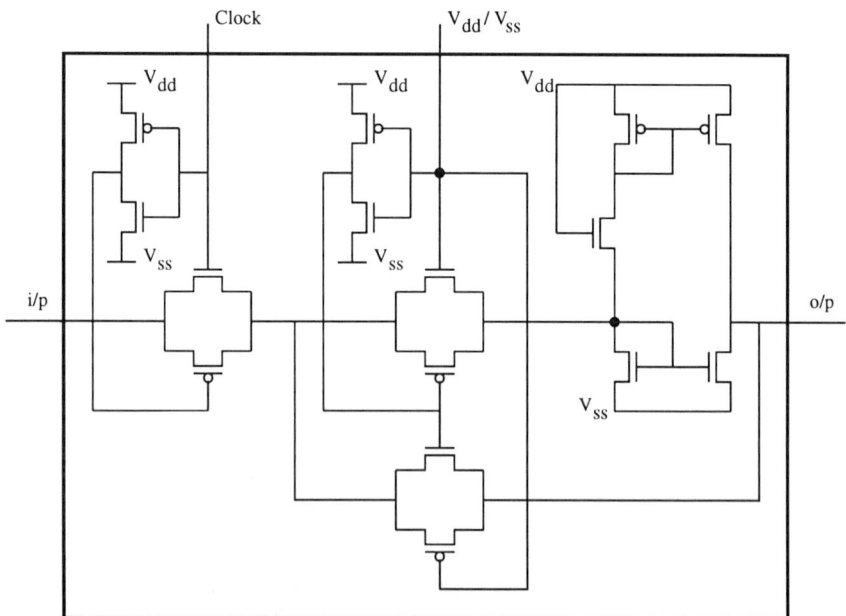

Figure 2.6 Circuit diagram of the developed SR programmable synapse

A Sampled-Data VLSI ANN Recognition System 53

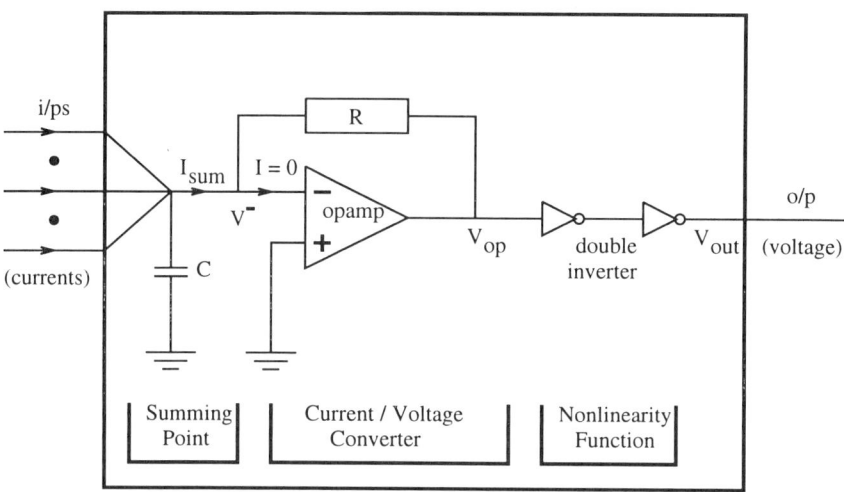

Figure 2.7 Schematic diagram of the developed CMOS analog neuron

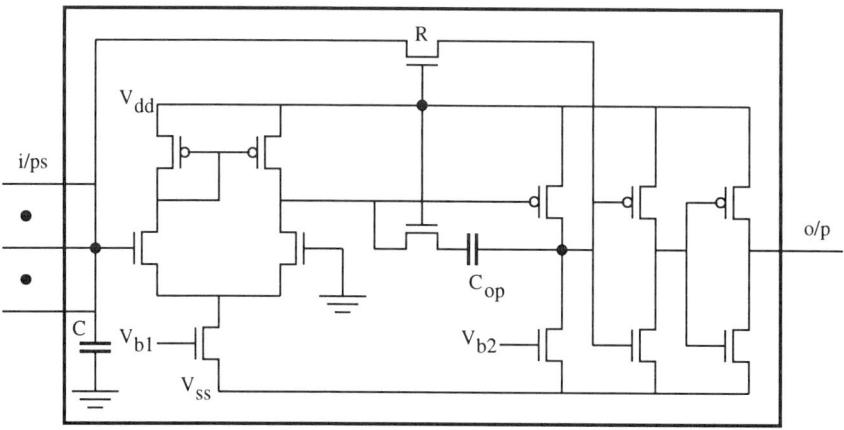

Figure 2.8 Circuit diagram of the developed CMOS analog neuron

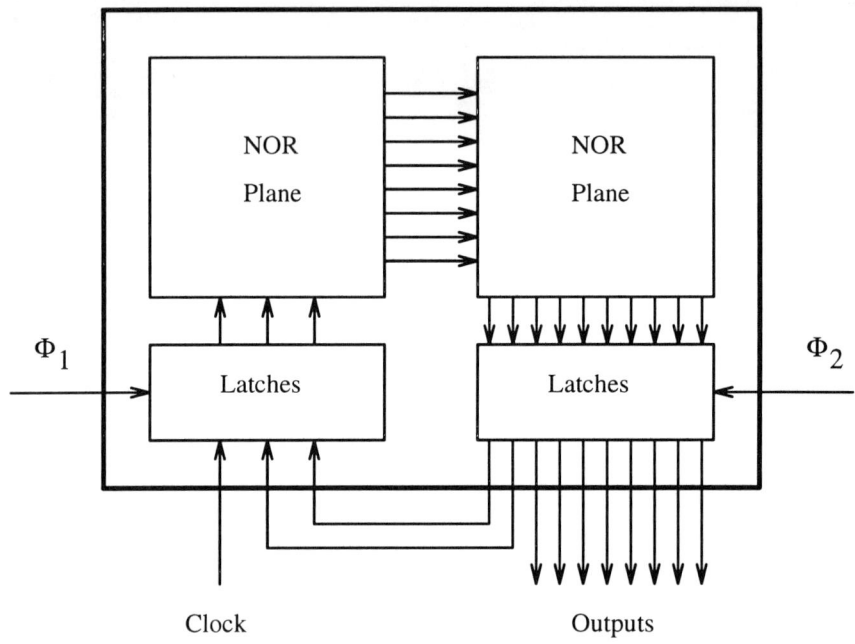

Figure 2.9 The block diagram of the designed PLA

Table 2.1 State Transition Table for the Designed PLA

Inputs			Outputs								
Clock	Present State		Next State		Output						
	B	A	B	A	1	2	3	4	5	6	7
0	0	0	0	0	1	1	1	1	1	1	1
1	0	0	0	1	0	1	1	1	1	1	1
0	0	1	0	1	0	0	1	1	1	1	1
1	0	1	1	0	0	0	0	1	1	1	1
0	1	0	1	0	0	0	0	0	1	1	1
1	1	0	1	1	0	0	0	0	0	1	1
0	1	1	1	1	0	0	0	0	0	0	1
1	1	1	0	0	0	0	0	0	0	0	0

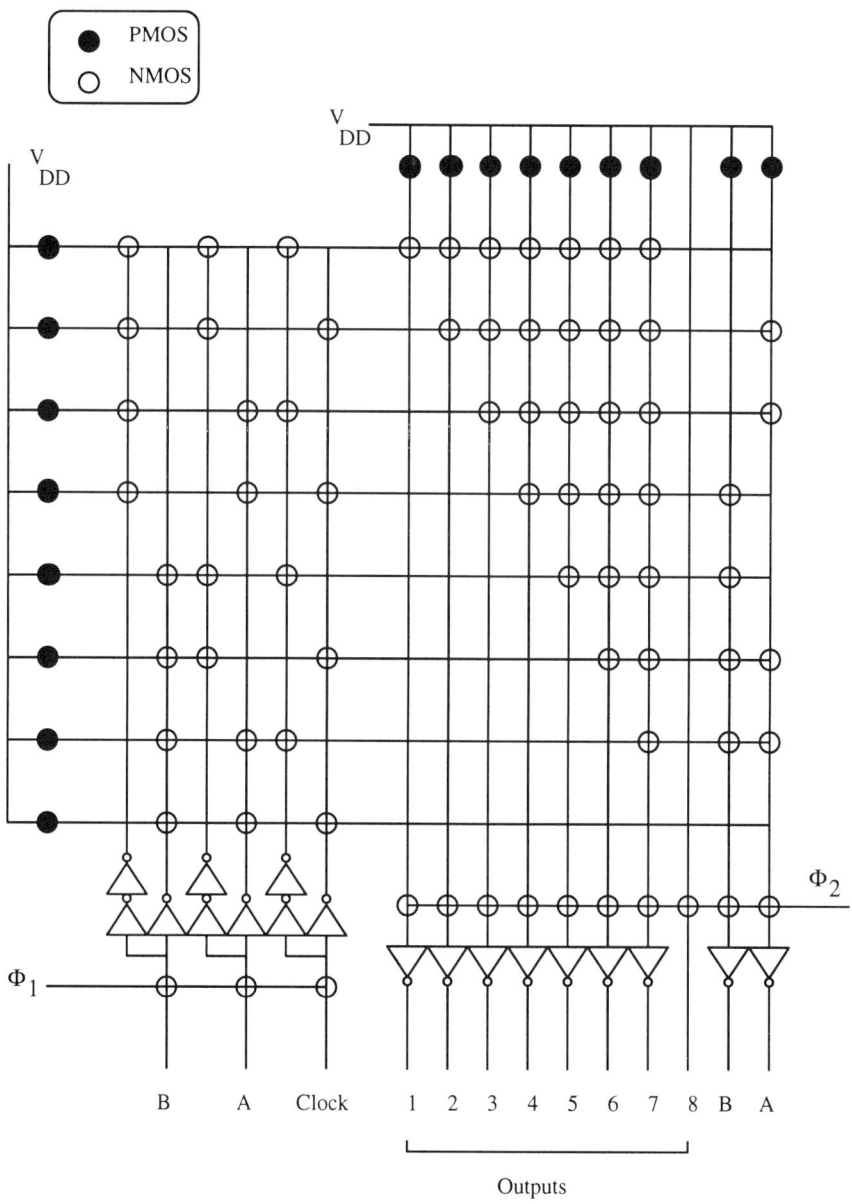

Figure 2.10 The circuit of the designed PLA

2.3.4 The ANN VLSI Testing Chip

A VLSI chip has been implemented in a 1.2 μm CMOS technology to test the developed programmable SR synapse and the simple analog neuron.

The layouts of the SR synapse and the analog neuron are shown in Figure 2.11 and in Figure 2.12, respectively.

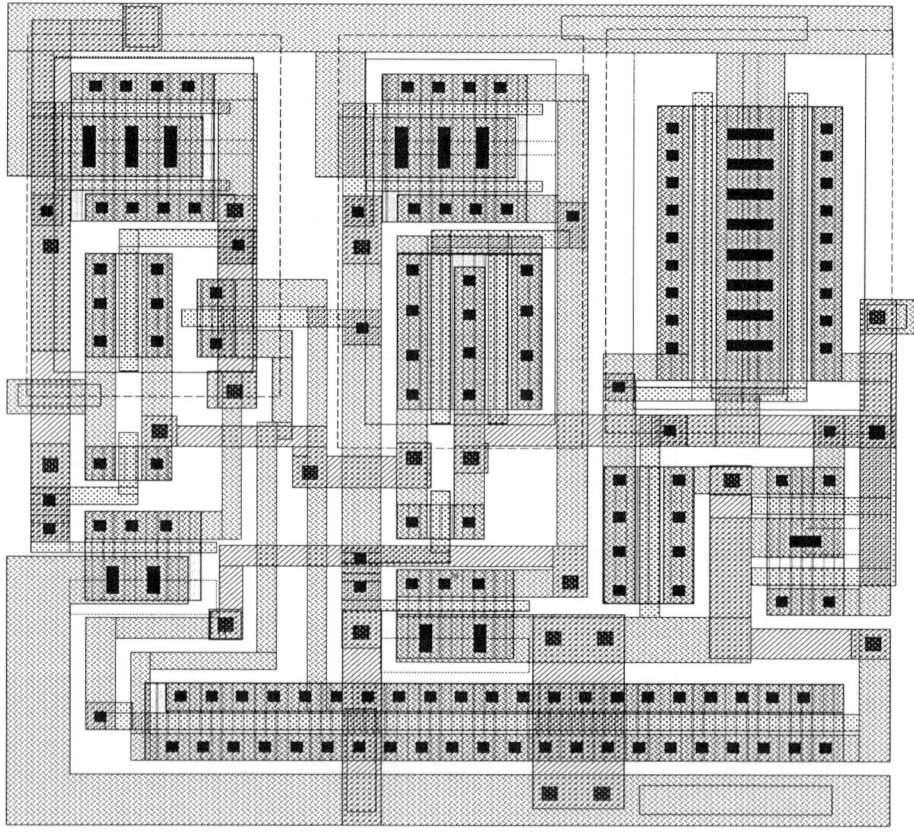

Figure 2.11 The layout of the developed SR programmable synapse

A Sampled-Data VLSI ANN Recognition System

The area consumed by the synapse is $120 * 120 \mu m^2$ while the area consumed by one neuron is $260 * 120 \mu m^2$.

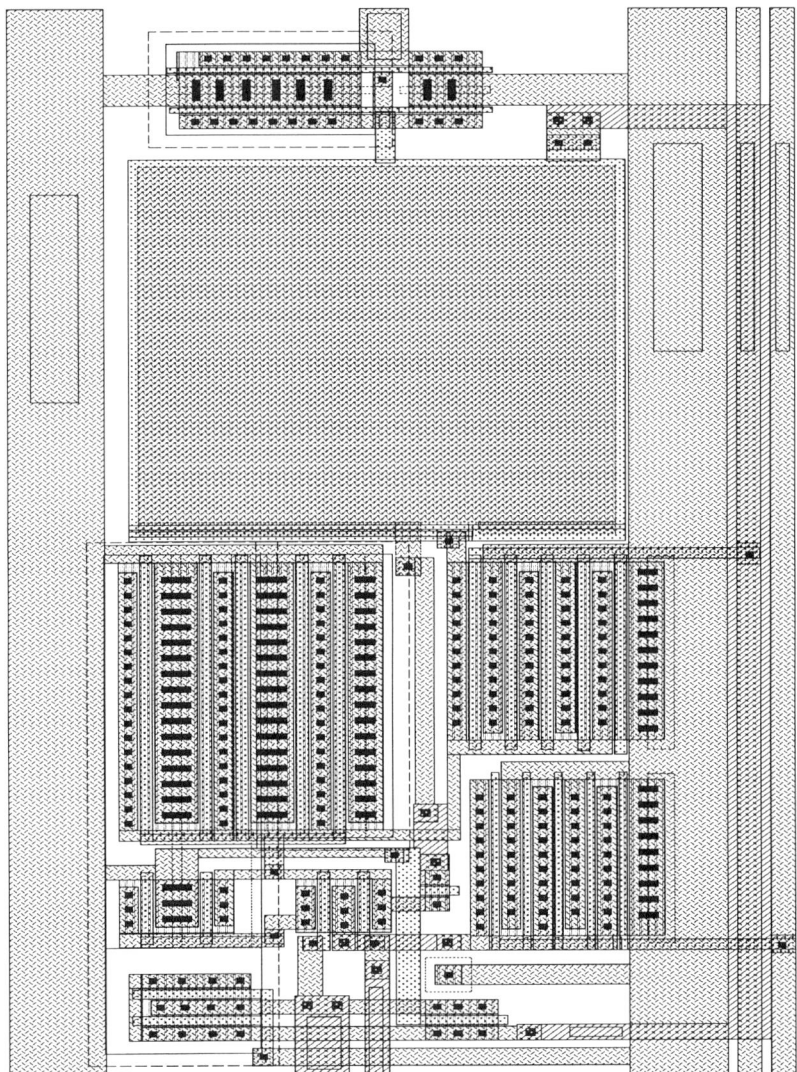

Figure 2.12 The layout of the developed CMOS analog neuron

The operation of the developed SR synapse was tested by observing the linear relation between the average conductance of the synapse and the duty-cycle of the controlling clock signal. Figure 2.13 shows close agreement between HSPICE simulation and experimental results of the implemented SR synapse.

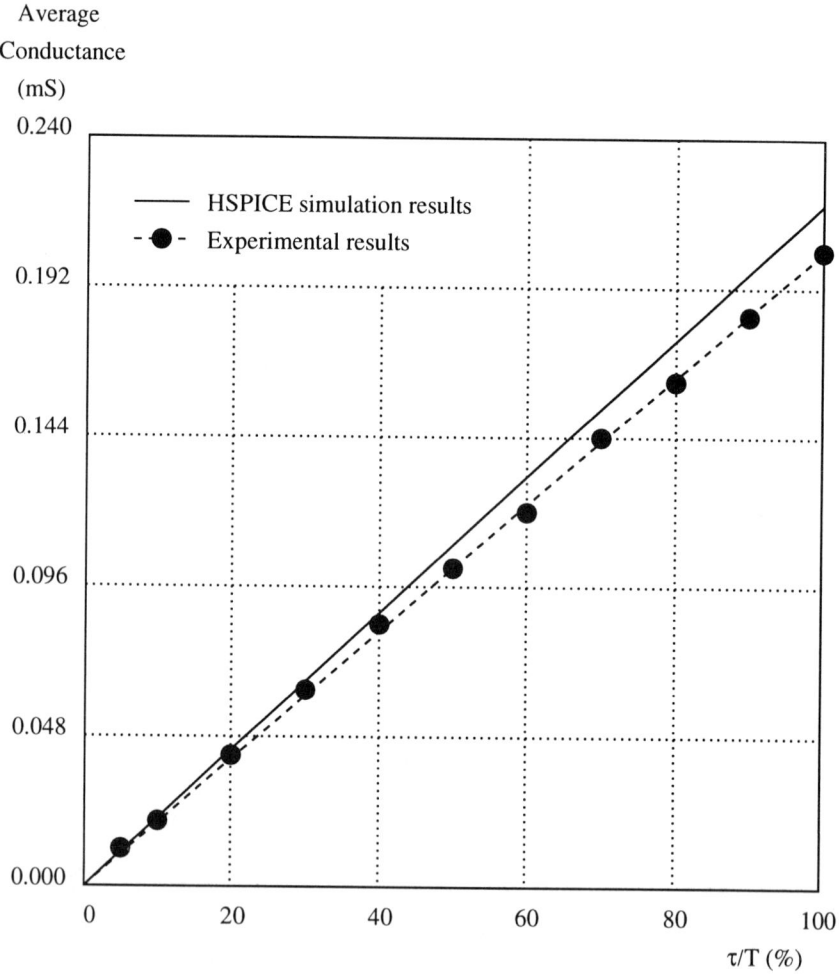

Figure 2.13 The linear relation between the average conductance of the SR synapse and the duty-cycle of the controlling clock signal

A Sampled-Data VLSI ANN Recognition System

The operation of the developed neuron was tested by observing the voltages at nodes V^-, V_{op}, and V_{out} (refer to Figure 2.7 on page 53) while varying the input current to the summing point of the opamp. The voltage at node V^- should be zero for any input current. The voltage at node V_{op} should be linearly dependent on the input current while the voltage at node V_{out} should have the tanh shape. Figure 2.14 shows both HSPICE simulation and experimental measurements of the three testing nodes in the implemented analog neuron.

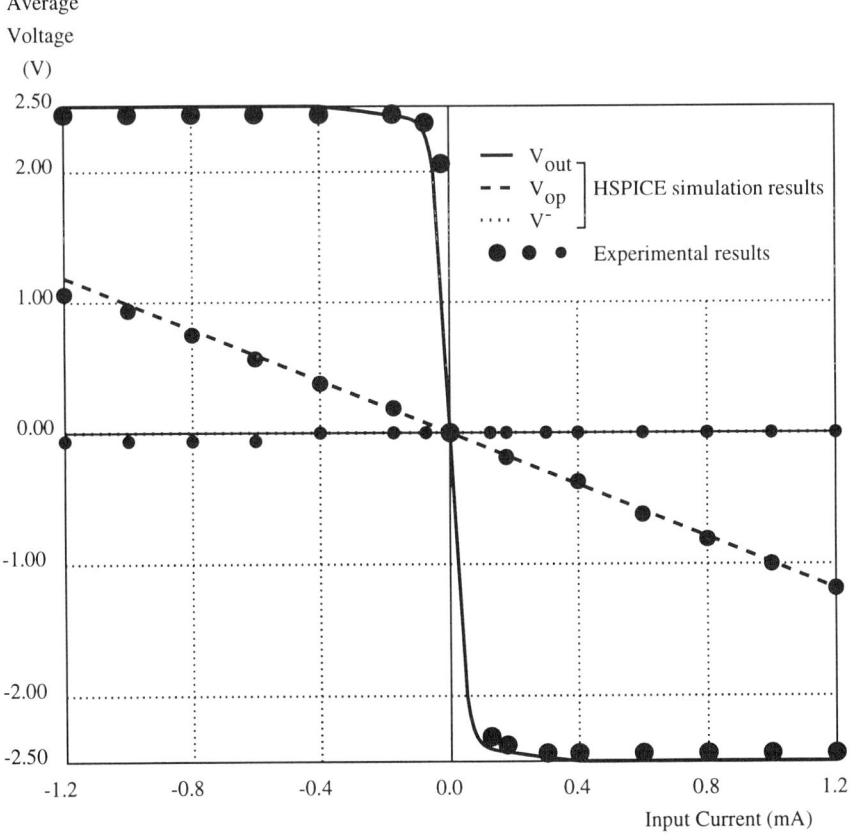

Figure 2.14 The voltages at the testing nodes of the CMOS analog neuron at different input summing current

2.4 THE PROTOTYPE MLP ANN MODEL ARCHITECTURE

Figure 2.15 shows an MLP ANN model architecture which is proposed in [47] to discriminate between two characters (e.g. T and C) independent of translation. In this MLP ANN model, input patterns are conceptualized as two-dimensional patterns superimposed on a square grid (e.g. $5 * 5$) called input field.

The hidden neurons are also organized into a two-dimensional grid with the same dimensions as the input field where each neuron is connected to only a small two-dimensional region of the input field (e.g. $3 * 3$) centered around the input node in the same location as the hidden neuron.

One output neuron is used to represent how closely an input pattern belongs to either one of the two characters. If the input pattern is a T (at any location), the output neuron has a positive value near "1" and if the input pattern is a C, the output neuron has a negative value near "-1".

To encode the translation invariance into the model architecture, all hidden neurons are constrained to learn exactly the same pattern of weights. In this way, the whole field of hidden neurons consists of simple replications of a single feature detector centered on different nodes of the input space. Also, the output neuron is connected to all hidden neurons through equal weights. These constraints guarantee translation independence and avoid any possible edge effects in the learning.

To use the above architecture for *multi*-character recognition, one output neuron will be needed for every different input character. This modification alleviates the need for a decoding stage after the ANN output stage to immediately identify the input character.

Each output neuron will be connected to all hidden neurons through a different set of equal weights. In addition, each output neuron will be used to represent how closely an input pattern belongs to one of the different input characters (If the input pattern is a T, the output neuron "T" has a positive value near "1" while all other output neurons have negative values near "-1").

Initial model simulations showed that this architecture can only be used for *two*-character recognition problems. To increase the power of the MLP ANN model architecture to recognize more characters, extra hidden planes (with the same dimensions as the existing hidden plane) are added.

A Sampled-Data VLSI ANN Recognition System

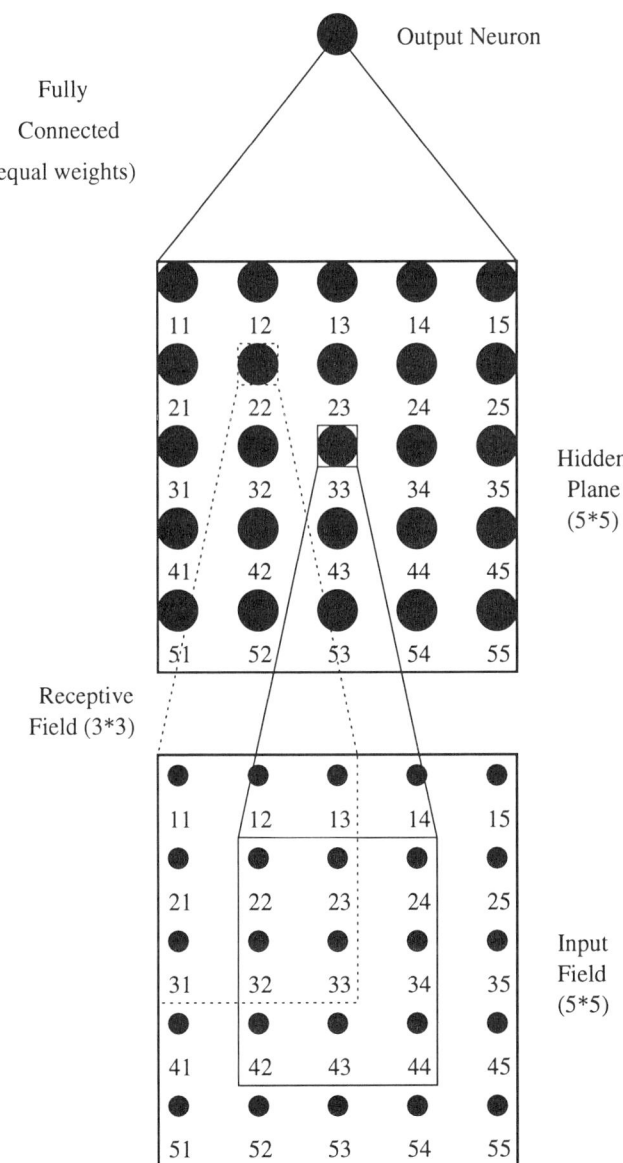

Figure 2.15 The MLP ANN model architecture

Each hidden neuron in the same location in any of the hidden planes is connected to the same group of nodes in the input field through different receptive fields.

The modified modular MLP ANN model architecture for *multi*-character recognition problems is shown in Figure 2.16.

2.5 MLP ANN MODEL SIMULATIONS

An MLP ANN model simulator which encodes the standard BP learning algorithm was developed. This developed MLP model simulator was modified to accommodate both the prototype model architecture and the developed circuits.

To accommodate the prototype model architecture, all weights which connect each output neuron to all hidden neurons are restricted to have the same value. Also, each hidden neuron is restricted to be connected to only a small two-dimensional region of the input field (e.g. $3*3$) centered around the input node in the same location as the hidden neuron. At the same time, every hidden neuron is restricted to be connected to the same pattern of weights (receptive field). To avoid the edge effects in the learning, the last row (column) in the input field is considered adjacent to the first row (column).

The modifications which accommodate the developed circuits can be summarized in the following points:

- The nonlinearity of the developed analog neuron circuit was encoded in the ANN model simulator rather than the ideal tanh function.
- During learning, the values of the weights were constrained to be less than the maximum value which can be achieved using the developed SR synapse.

The modified ANN simulator was used to perform software simulations on different recognition problems using the MLP ANN model architecture shown in Figure 2.16. Four different recognition problems were selected as benchmark problems to test the functionality of the modified model architecture. These recognition problems are explained in detail in the following subsections.

A Sampled-Data VLSI ANN Recognition System

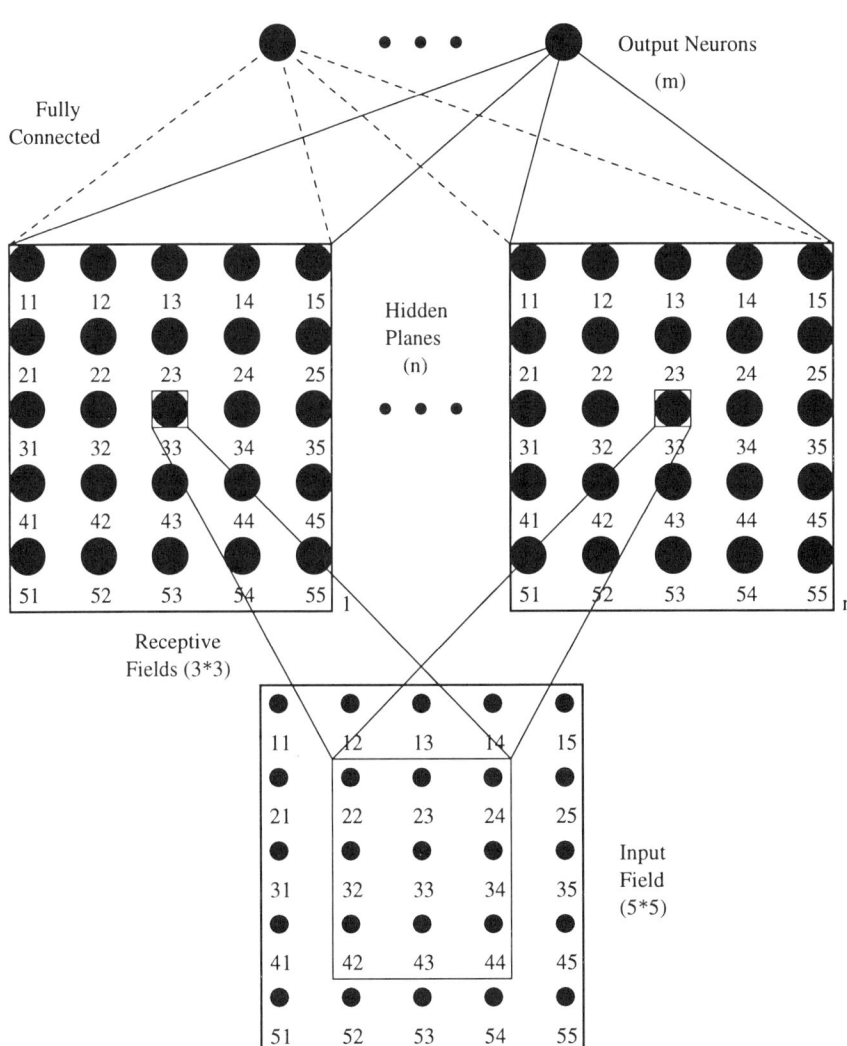

Figure 2.16 The proposed multi-character MLP ANN model architecture

2.5.1 The Two-Character Recognition Problem

This benchmark problem is a two-character recognition problem where the input patterns are restricted to be the two characters (T and C) and the testing patterns are restricted to be one of the original two characters (T and C) distorted by a limited percentage of random noise.

This two-character recognition problem is selected as a prototype to demonstrate the next steps of the implementation procedure. The two characters (T and C) are represented as $3*3$ pixels with only one pixel difference between the two characters. The input field is selected to be only $3*3$ to minimize the area of the prototype ANN chip. To avoid the edge effects in the learning, the third row (column) in the input field is considered adjacent to the first row (column). Each hidden neuron is connected to the input nodes through a receptive field of $3*3$.

ANN model simulations for this prototype MLP ANN model, which distinguishes between the two characters T and C, were conducted using the modified ANN simulator. Only one hidden plane was needed to perform the required recognition task using the prototype MLP ANN model architecture with a single output neuron.

In this benchmark problem, only the two characters T and C were used in the training process. 1254 iterations were needed to complete the learning process of this pattern recognition problem. A set of 100 weights is extracted from the simulation results. The maximum number of different weights for this prototype model is only 12 weights. The output neuron is connected to all hidden neurons through 9 equal weights as well as to the threshold node through another weight (maximum of 2 different values). Each hidden neuron is connected through 10 weights to the 9 input nodes and the threshold node (maximum of 10 different values).

Figure 2.17 shows the symbols used to represent those 10 weights in the VLSI architectures presented in a subsequent section.

Different sets of 200 testing patterns (100 from each character) were extracted from the original two digits (T and C) after adding random noise up to a certain percentage of the original values. Figure 2.18 shows the MLP ANN model simulation results using noisy test patterns.

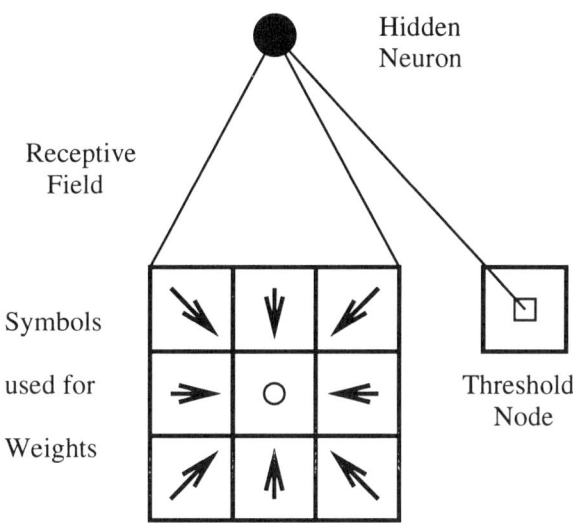

Figure 2.17 Representation of the symbols used for the receptive field's weights in the proposed VLSI architectures

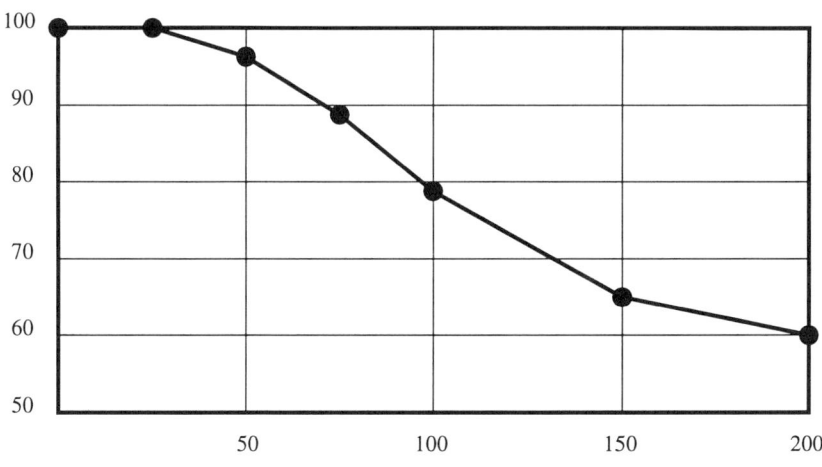

Figure 2.18 The MLP testing simulation results of the two-character recognition problem

It is worth mentioning that since there are only two possible outputs in this problem, the minimum percentage of correct testing patterns should be equal to 50% (the blind guess percentage).

2.5.2 Digit Recognition Problems

Three different digit recognition problems were selected as benchmark problems to test the functionality of the modified model architecture. The input digits used in these problems are shown in Figure 2.19. Each digit is represented as $3*5$ pixels superimposed on a $7*7$ input field.

The first benchmark problem is a two-digit recognition problem where the input patterns are restricted to be the two digits (0 and 1) and the testing patterns are restricted to be one of the original two digits (0 and 1) distorted by a limited percentage of random noise.

The second benchmark problem is a two-digit recognition problem where the input patterns are the original ten digits and the testing patterns can be any one of the original ten digits distorted by a limited percentage of random noise. Three output neurons are used in this case where the third neuron is added to represent any of the other digits (the reject neuron).

The third benchmark problem is the ten-digit recognition problem where ten output neurons are used and the input patterns are the original ten digits. The testing patterns can be any one of the original ten digits distorted by a limited percentage of random noise.

2.5.2.1 The 0/1 recognition problem:

In this benchmark problem, only the two digits "0" and "1" were used in the training process. The used MLP ANN model architecture had two output neurons and only one hidden plane was needed to perform the required recognition task. The receptive field for each hidden neuron was selected to be $3*3$ to minimize the number of synapses in the ANN model.

1295 iterations were needed to complete the learning process of this pattern recognition problem. The set of weights are extracted after learning and used in the testing phase.

A Sampled-Data VLSI ANN Recognition System

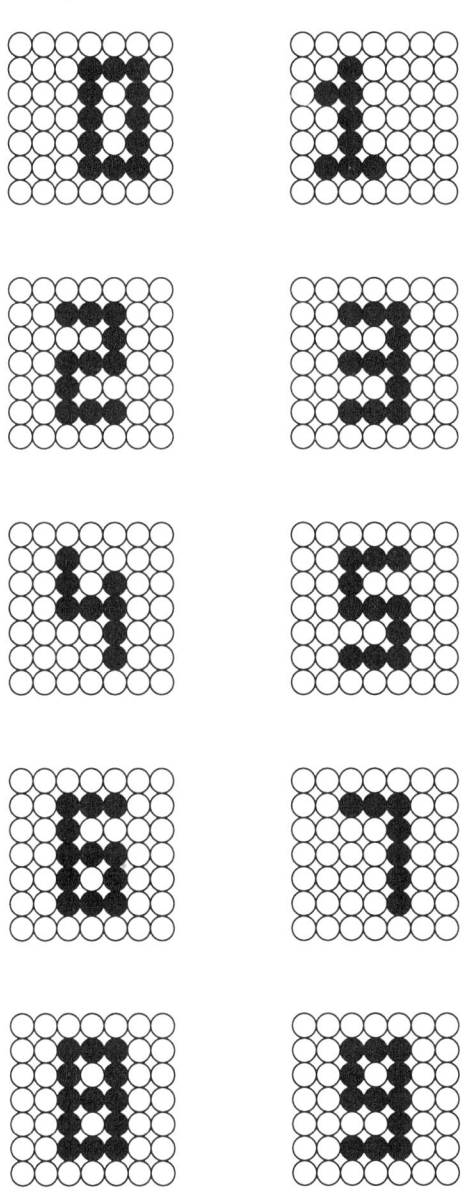

Figure 2.19 The input digits used in the ANN software simulations

Figure 2.20 shows the MLP ANN model simulation results using different sets of 200 testing patterns (100 from each digit) extracted from the original two digits (0 and 1) after adding random noise up to a certain percentage of the original values.

It is worth mentioning that since there are only two possible outputs in this problem, the minimum percentage of correct testing patterns should be equal to 50% (the blind guess percentage).

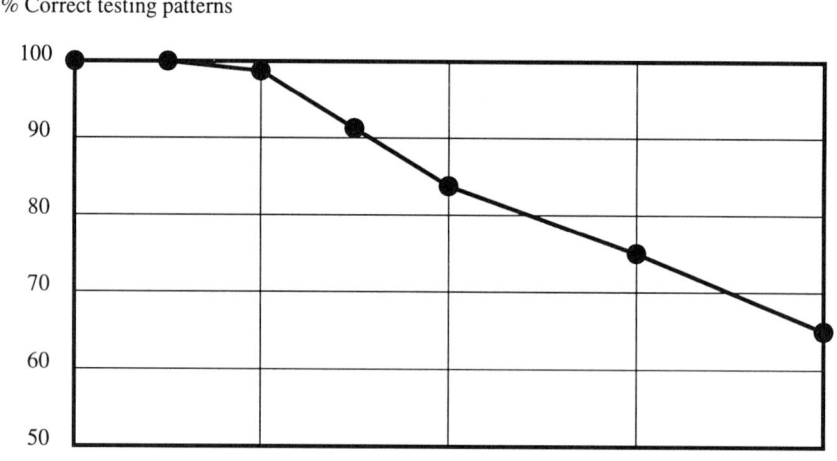

Figure 2.20 The MLP testing simulation results of the 0/1 recognition problem

2.5.2.2 *The 0/1/reject recognition problem:*

In this benchmark problem, all input digits were used in the training process. In the used MLP ANN model architecture, three output neurons and three hidden planes were needed to perform the required recognition task. The receptive field for each hidden neuron was selected to be 3 * 3 to minimize the number of synapses in the ANN model.

4056 iterations were needed to complete the learning process of this pattern recognition problem. The set of weights are extracted after learning and used in the testing phase.

Figure 2.21 shows the MLP ANN model simulation results using different sets of 1000 testing patterns (100 from each digit) extracted from the original ten digits after adding random noise up to a certain percentage of the original values.

Since there are three output neurons in this benchmark problem, the minimum percentage of correct testing patterns should be equal to 33.33% (the blind guess percentage).

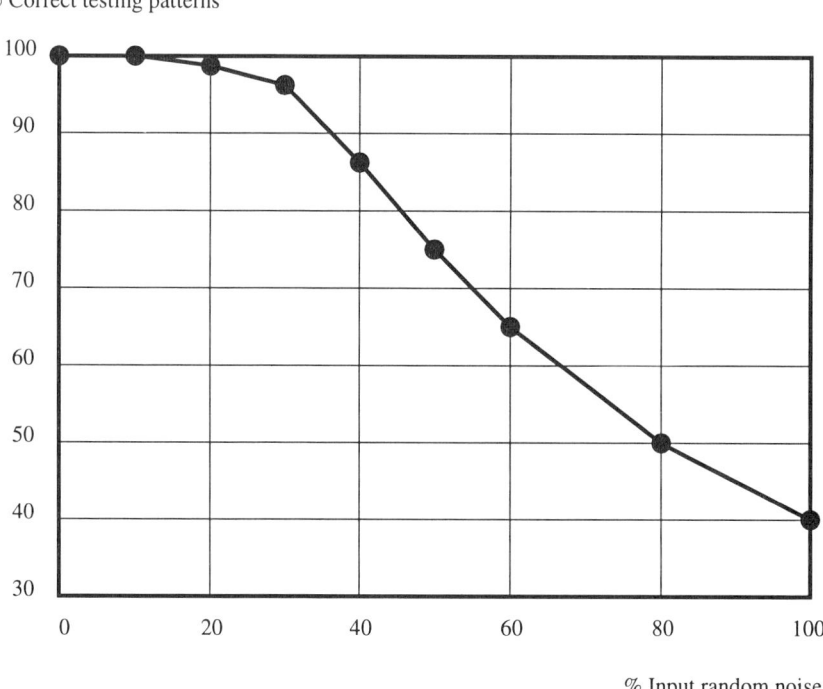

Figure 2.21 The MLP testing simulation results of the 0/1/reject recognition problem

2.5.2.3 The ten-digit recognition problem:

In this benchmark problem, all input digits were used in the training process. In the used MLP ANN model architecture, ten output neurons, three hidden planes, and a 4∗4 receptive field for each hidden neuron were needed to perform the required recognition task.

Two different techniques were used to perform the learning process of this pattern recognition problem. The first one is the block-learning technique which is the same technique used in previous problems (i.e. presenting all training patterns in one training process.

The second technique is an incremental-learning one in which a sequence of multiple training cycles is performed to solve the given pattern recognition problem. At the beginning, only two patterns are used to train the full MLP ANN model. After learning, the resulting set of weights is used as the initial set for the next training cycle in which three patterns (including the previous two) are used as input patterns. This procedure continues until all input patterns are used.

Using both techniques, two different sets of weights were extracted, after learning, and used in the testing phase. Different sets of 1000 testing patterns (100/digit) were extracted from the original 10 digits after adding random noise up to a certain percentage of the original vlaues. Figure 2.22 shows the MLP ANN model simulation results using these testing patterns.

Since there are ten output neurons in this benchmark problem, the minimum percentage of correct testing patterns should be equal to 10% (the blind guess percentage).

2.5.3 Overall Performance of the Proposed Architecture

Table 2.2 shows the relation between the number of hidden planes and the maximum number of digits that can be recognized using the proposed architecture for multi-character recognition problems.

A Sampled-Data VLSI ANN Recognition System

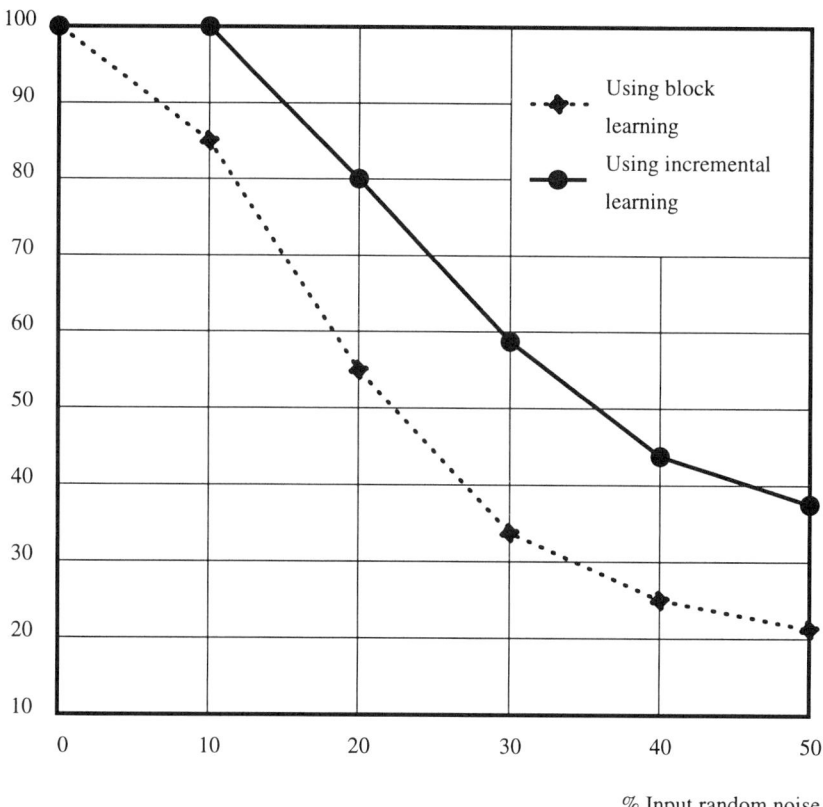

Figure 2.22 The MLP simulation results for testing patterns of the ten-digit recognition problem

Table 2.2 The relation between the number of hidden planes and the maximum number of recognized digits

Number of hidden planes	Max. number of recognized digits
1	2
2	6
3	18

2.6 ANN CIRCUIT SIMULATIONS

The prototype MLP ANN model architecture, which is described in the previous section to solve the two-character recognition problem, is implemented as an HSPICE input file using the developed ANN circuits. The weights extracted from the ANN model simulations (after learning) have been quantized to one of 100 divisions ($<$ 7 bit resolution), where the maximum value corresponds to the maximum allowable value of the weight; when using a 100% duty-cycle controlling clock signal. The duty-cycles of controlling signals have been calculated using the corresponding quantized weights and encoded in the HSPICE file.

HSPICE simulations of the ANN circuit, using fixed $1MHz$ clock signals, have been conducted for the two input patterns (T and C). Each input pattern was applied for $10\mu S$. The output neuron has a high positive output voltage when applying an input pattern T to the input nodes of the ANN circuit; while the output neuron has a high negative output voltage when applying an input pattern C. The output signal of the ANN circuit, for both cases of inputs ($20\mu S$), is shown in Figure 2.23.

It can be noticed from this Figure that the output neuron of the prototype two-layer MLP needs only $2\mu S$ (2 clock cycles) to correctly recognize the applied input pattern. This result is in agreement with the results reported in [50] for a two-layer MLP ANN circuit used to solve the XOR problem. Therefore, it can be concluded that circuit simulation results confirm the validity of the ANN circuit to perform the desired recognition task.

2.7 THE DEVELOPED VLSI ARCHITECTURES

2.7.1 The Parallel VLSI Architecture

The parallel VLSI architecture, shown in Figure 2.24, can be used to implement the prototype MLP model. In this parallel architecture, the input nodes are organized in a vertical array while the hidden neurons are organized in a horizontal array. Both input nodes and hidden neurons are represented by circles where the number written beside every circle represents the position of the nearest node/neuron in the two-dimensional space of the MLP ANN model architecture (refer to Figure 2.15). The connections between the hidden neurons and the input nodes are shown where the synapses with the same value are

A Sampled-Data VLSI ANN Recognition System

represented by the same symbol. The output neuron is connected to all hidden neurons through synapses with the same value.

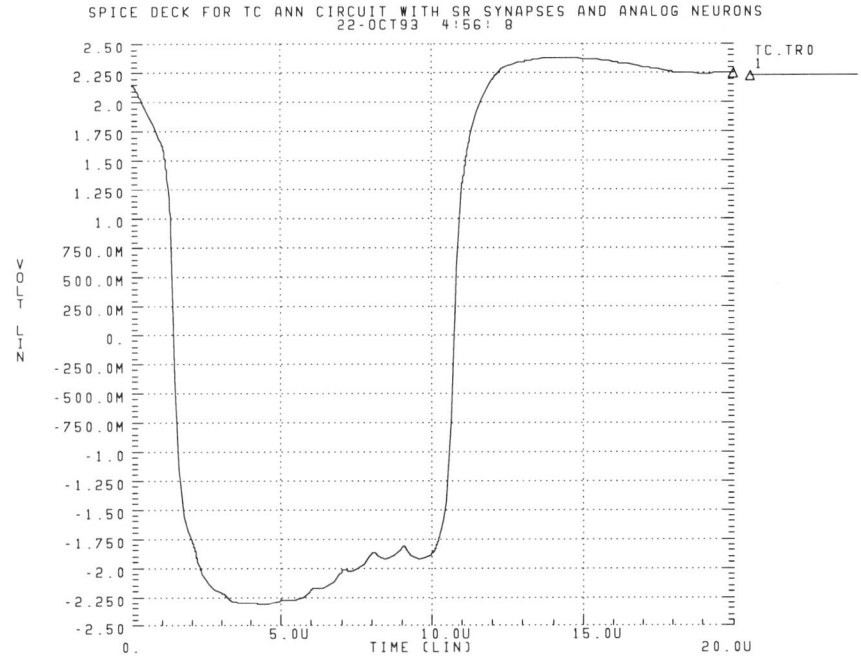

Figure 2.23 The output signal of the prototype MLP circuit using HSPICE

This parallel VLSI architecture needs a technology with three metal layers to efficiently implement the MLP model. Two metal layers are required to implement the connections between the hidden neurons and the input nodes while the third one is needed to connect the controlling clock signals to their corresponding synapses. To overcome the problem of having a technology of only two layers of metal, all vertical metal connections are restricted to be of one type of metal while the other type is used for horizontal connections.

The parallel VLSI architecture is suitable for implementing the fully-connected synapses matrix between the output and the hidden neurons. For character recognition problems with larger input field, the parallel VLSI architecture, which implements the partially-connected synapses matrix between the hidden neurons and the input nodes, consumes larger area than what is actually required. That is because the synapses matrix needs to be expanded in both directions to accommodate the increase in the number of both input nodes and hidden neurons, although each hidden neuron is only connected to a fraction of the input nodes.

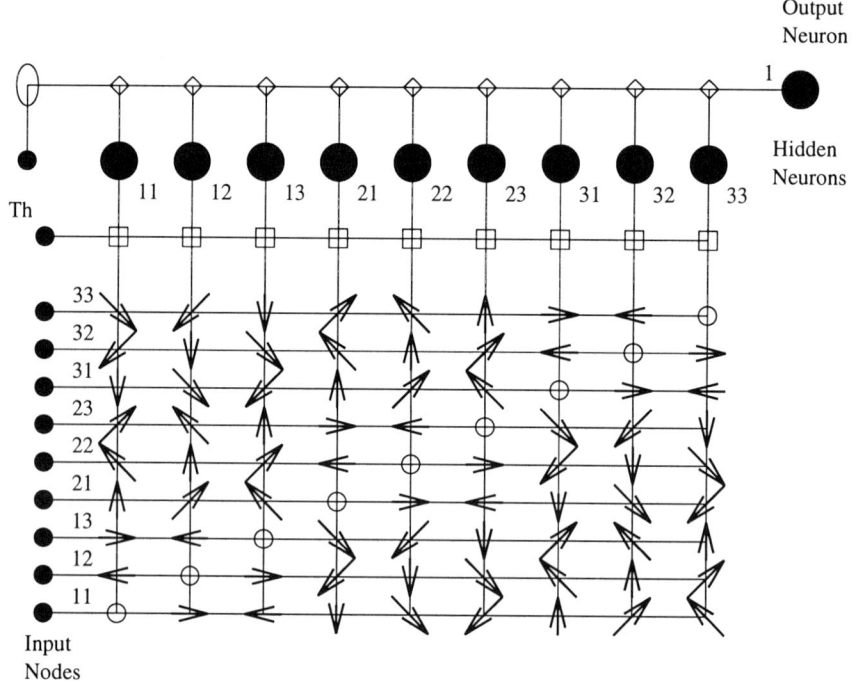

Figure 2.24 The parallel VLSI architecture

2.7.2 The Diamond VLSI Architecture

For large input fields ($> 4 * 4$), the developed diamond architecture, shown in Figure 2.25, is a a better VLSI implementation for the partially-connected synapses matrix. In this architecture, both the input nodes and the hidden neurons are organized horizontally. The main advantage of this architecture with respect to the parallel architecture is that the number of rows in the synapses matrix is fixed and does not increase with the increase in the input field.

A Sampled-Data VLSI ANN Recognition System

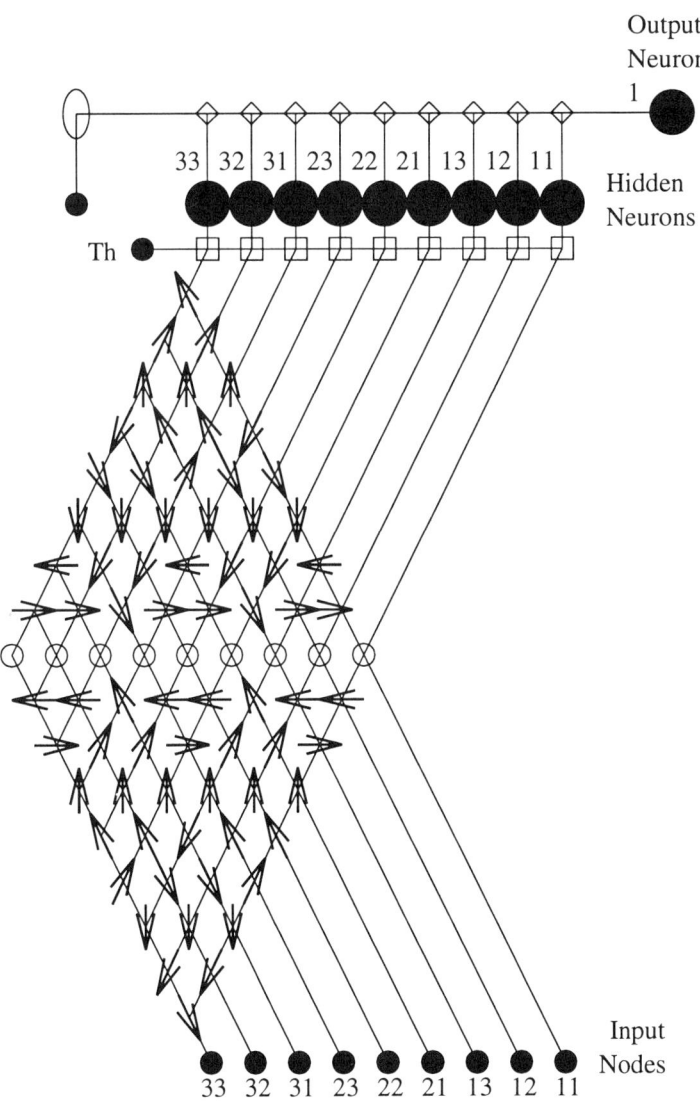

Figure 2.25 The diamond VLSI architecture

2.7.3 The Parallelogram VLSI Architecture

The number of rows in the partially-connected synapses matrix can be reduced to its minimum value (the number of input nodes connected to one hidden neuron) using the parallelogram VLSI architecture shown in Figure 2.26 (where input nodes are repeated only for clarity). It is worth mentioning that the routing of the controlling clock signals' metal connections is much easier and consumes a smaller area using the developed parallelogram VLSI architecture.

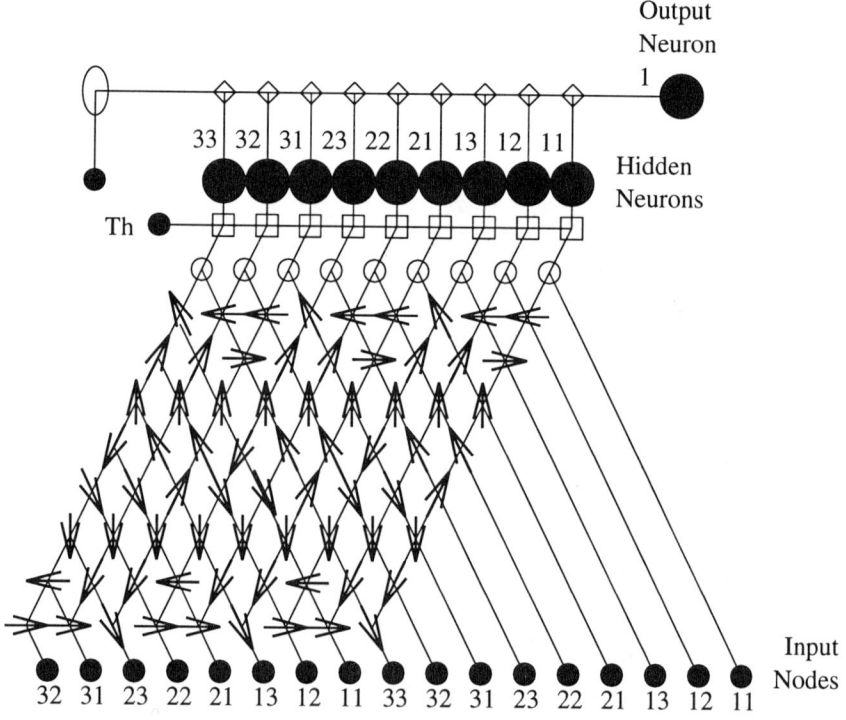

Figure 2.26 The developed parallelogram VLSI architecture

For a character recognition problem with $n*m$ input field and $k*l$ receptive field, the ratioed area consumed in the partially-connected synapses matrix is proportional to $(n*m)^2$ for the parallel VLSI architecture. For the diamond VLSI architecture, the area is proportional to $(2(k*l) - 1)(n*m)$ while it is proportional to $(k*l)(n*m)$ for the parallelogram VLSI architecture.

A Sampled-Data VLSI ANN Recognition System

Table 2.3 A comparison between the three proposed VLSI architectures in the ratioed area consumed in the synapses' matrix

Dimensions of input field	Parallel VLSI architecture	Diamond VLSI architecture	Parallelogram VLSI architecture
3 * 3	9 * 9	17 * 9	9 * 9
4 * 4	16 * 16	17 * 16	9 * 16
5 * 5	25 * 25	17 * 25	9 * 25
10 * 10	100 * 100	17 * 100	9 * 100

Table 2.3 shows a comparison between the three VLSI architectures in the ratioed area consumed in the partially-connected synapses matrix for fixed receptive field (3 * 3) and different input fields.

2.8 THE DEVELOPED TWO-CHARACTER ANN RECOGNIZER

A SR VLSI ANN chip is designed to implement the MLP circuit which can be used to distinguish between two characters (*two*-character recognizer). The layout of the developed *two*-character ANN recognizer chip is shown in Figure 2.27. The chip consumes an area of $3400 * 4600 \mu m^2$ using a 1.2 μm CMOS technology.

The organization of the 100 synapses and the 10 neurons implemented on the ANN chip can be described as follows:

- A 9*9 synapses matrix, which represents the synapses connecting the input nodes to the hidden neurons, implemented using the developed parallelogram VLSI architecture;
- 9 synapses, which connect the input threshold node to the hidden neurons, organized as a row;
- 9 hidden neurons organized as a row in the middle of the chip;
- 1 * 9 synapses vector which represents the synapses connecting the hidden neurons to the output neuron;

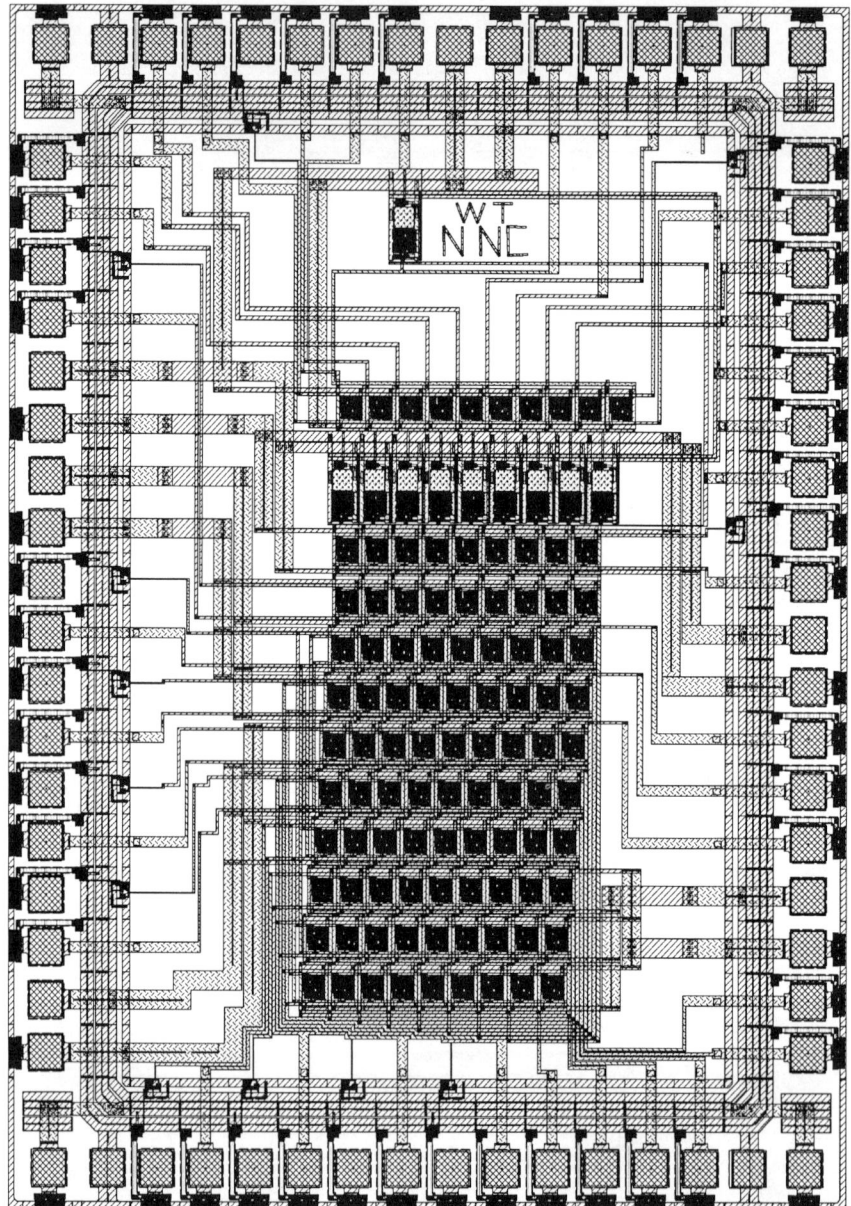

Figure 2.27 The layout of the developed two-character ANN recognizer chip

- 1 synapse which connects the hidden threshold node to the output neuron; and
- 1 output neuron.

The SR ANN chip has 68 pads which can be itemized as follows:

- 9 for input nodes;
- 2 for threshold nodes;
- 1 for output neuron;
- 9 for hidden neurons;
- 12 for the controlling clock signals which determine the magnitudes of the weights;
- 12 for the voltage signals which determine the signs of the weights;
- 3 for reference voltage nodes needed for the operation of the neurons; and
- The remaining pads for supplying both the ANN circuit and the pads with the two power supply voltage sources.

HSPICE simulations have been conducted for the extracted circuit of the designed ANN chip to demonstrate the functionality of the chip to perform the required recognition task.

2.9 THE PROPOSED MULTI-CHARACTER ANN RECOGNITION SYSTEM

The proposed architecture for the multi-chip multi-character ANN recognition system is shown in Figure 2.28. For an n-character ANN recognition system, the proposed architecture consists of a number of modular ANN recognizer chips and one output-stage ANN chip. The input voltages, which represent the input pattern, should be available to all the modular ANN recognizer chips in the multi-character ANN system.

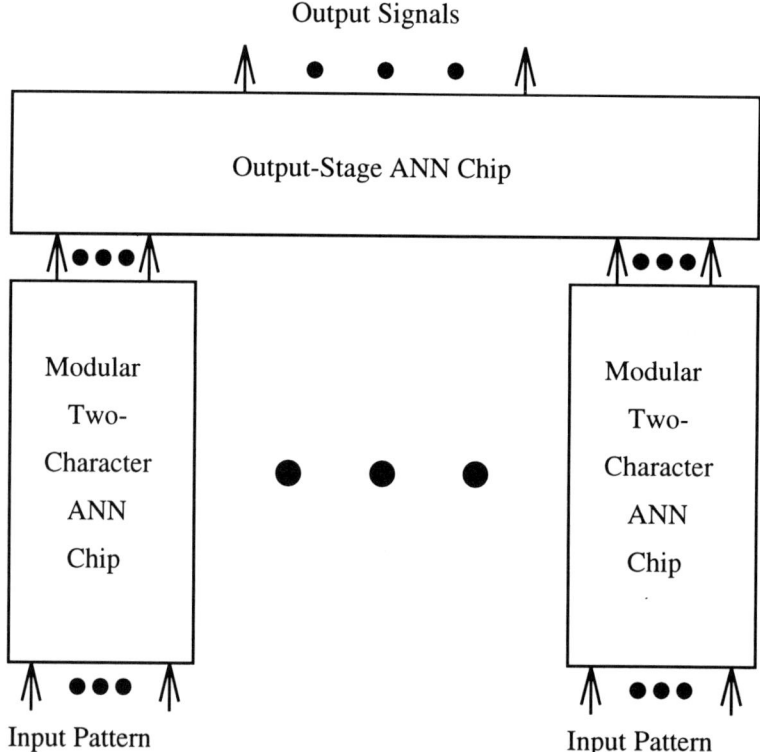

Figure 2.28 The proposed architecture for multi-character ANN recognition systems

2.9.1 The Modular ANN Recognizer

The modular ANN recognizer is a modified version of the developed two-character ANN recognizer chip. The number of the modular ANN recognizer chips should be equal to the number of hidden planes required to solve the n-character recognition problem. This means that the hidden neurons in a modular ANN chip represent one hidden plane (refer to Figure 2.16 on page 63).

The modular chip, shown in Figure 2.29, is the developed two-character ANN recognizer chip with the following modifications:

- The output neurons are removed (they are included in the output-stage ANN chip).
- The number of rows of the output synapses should be equal to the number of output neurons, n (i.e. there will be one row in the synapses matrix between the hidden units and the output neurons for every output neuron).

2.9.2 The Output-Stage ANN Chip

The output-stage ANN chip, shown in Figure 2.30, is needed to sum the corresponding output currents from all modular ANN recognizer chips and then apply the summed currents to the output neurons to give the final output signals.

2.10 CONCLUSIONS

The SR implementation of ANN captures both the advantages of analog implementation (i.e. parallelism, compactness, and simplicity) and the programmability of digital implementation. Experimental results of a fabricated ANN chip, which implements novel programmable SR synapses and a simple CMOS analog neuron, demonstrate the feasibility of an SR CMOS VLSI implementation of ANNs.

An extended MLP model architecture is proposed for use in solving *multi-*character recognition problems. HSPICE simulations have been conducted to demonstrate the functionality of the SR CMOS ANN circuit to perform the required recognition task.

The parallelogram VLSI architecture developed to implement the partially-connected synapses matrices is regular, modular, and easy to expand.

A novel SR ANN chip is designed to implement a *two*-character recognizer. A multi-chip multi-character ANN recognition system is proposed. These chips are implemented using a 1.2 μm CMOS technology. HSPICE simulations of the extracted circuits show that the developed chips are capable of performing the required recognition task.

The implemented SR ANN circuits, using the developed building blocks, exhibit a very high modularity and can be considered as a good member in the family of the mixed-mode VLSI ANNs.

A Sampled-Data VLSI ANN Recognition System

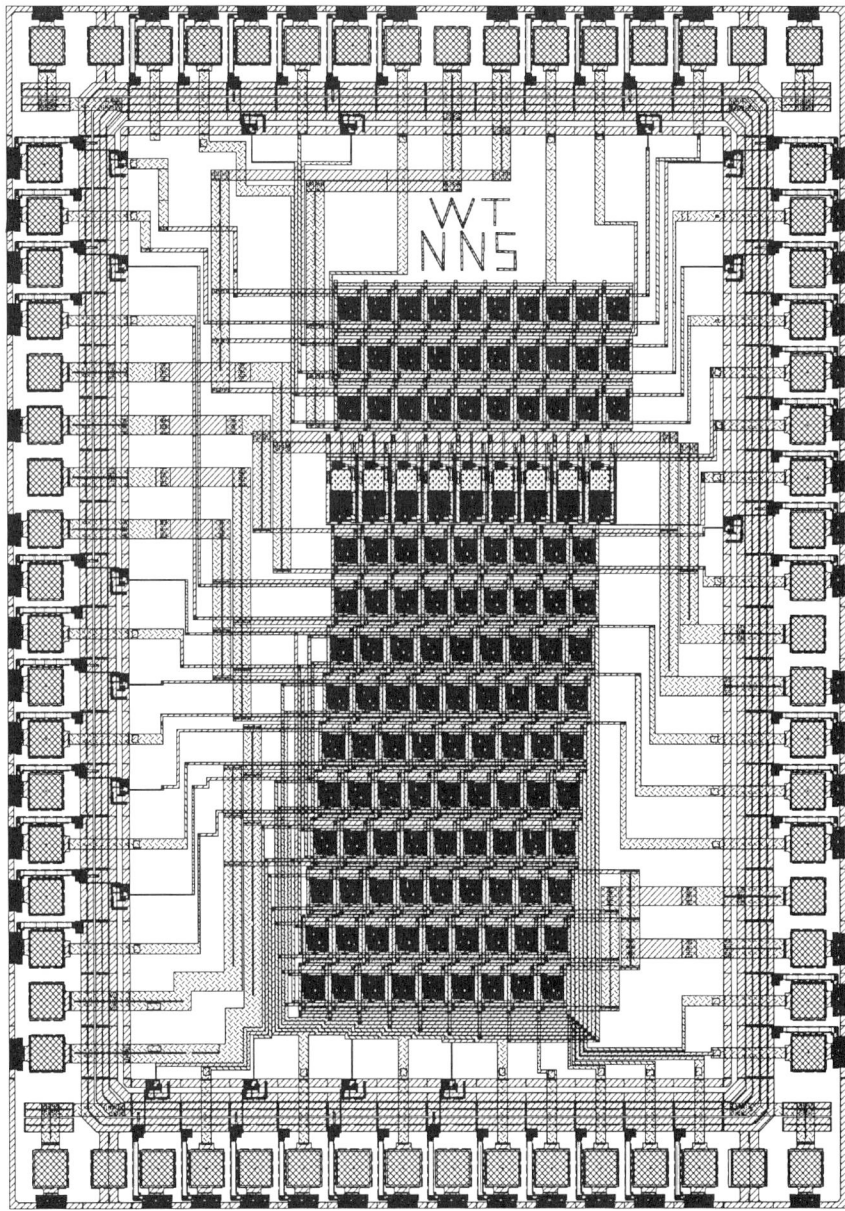

Figure 2.29 The layout of the modular ANN recognizer chip (3 outputs)

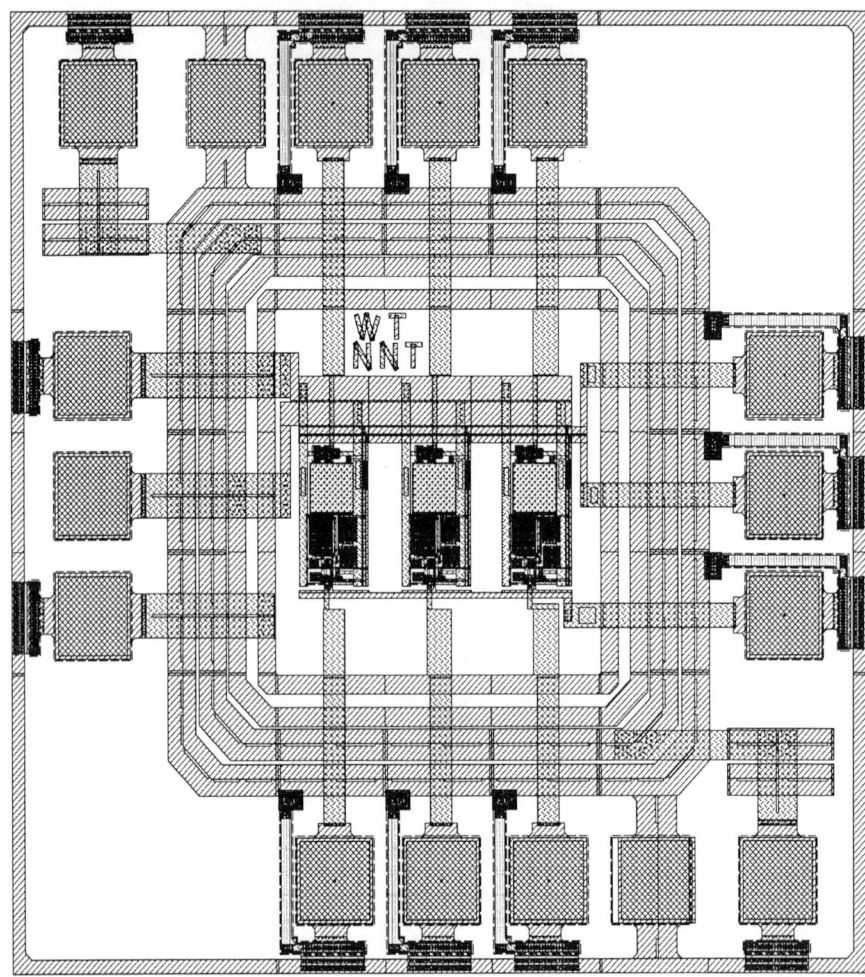

Figure 2.30 The layout of the developed output-stage ANN chip (3 outputs)

REFERENCES

[1] Bernstein, J., *"Profiles: Marvin Minsky,"* The New Yorker, pp.50-126, December 1981.

[2] Hay, J., Martin, F., and Wightman, C., *"The MARK I Perceptron, Design and Performance,"* IRE International Convention Record: Part 2, New York: IRE, pp.78-87, 1960.

[3] Widrow, B., and Hoff, M., *"Adaptive Switching Circuits,"* WESCON Convention Record: Part 4, pp.96-104, 1960.

[4] Crane, H., *"Neuristor: A Novel Device and System Concept,"* Proceedings of the IRE, 50, pp.2048-2060, 1962.

[5] Hecht-Nielsen, R., *"Performance of Optical, Electro-Optical, and Electronic Neurocomputers,"* Proceedings of the SPIE, 634, pp.277-306, 1986.

[6] Farhat, N., Psaltis, D., Prata, A., and Paek, E., *"Optical Implementation of the Hopfield Model,"* Applied Optics, 24, pp.1469-1475, 1985.

[7] Denker, J. (Ed.), *"AIP Conference Proceedings 151: Neural Networks for Computing,"* New York: American Institute of Physics, 1986.

[8] Graf, H. P., Jackel, L., and Hubbard, W., *"VLSI Implementation of a Neural Network Model,"* IEEE Computer, Volume 21, Number 3, pp.41-49, March 1988.

[9] Graf, H. P., et al., *"A CMOS Associative Memory Chip,"* Proceedings of IJCNN, Volume III, pp.461-468, 1987.

[10] Jackel, L., et al., *"Artificial Neural Networks for Computing,"* Journal of Vacuum Science & Technology B (Microelectronics: Processing and Phenomena), Volume 4, Number 1, pp.61-63, January/February 1986.

[11] Mead, C., and Mahowald, M., *"A Silicon Model of Early Visual Processing,"* Neural Networks, 1, pp.91-97, 1988.

[12] Paulos, J. J., and Hollis, P. W., *"Neural Networks using Analog Multipliers,"* Proceedings of IEEE ISCAS, Volume I, pp.499-502, 1988.

[13] Verleysen, M., et al., *"Neural Networks for High-Storage Content-Addressable Memory: VLSI Circuit and Learning Algorithm,"* IEEE Journal of Solid State Circuits, Volume 24, Number 3, pp.562-569, June 1989.

[14] Rossetto, O., et al., *"Analog VLSI Synaptic Matrices as Building Blocks for Neural Networks,"* IEEE Micro, Volume 9, Number 6, pp.56-63, December 1989.

[15] Jou, I. C., Wu, C., and Liu, R., *"Programmable SC Neural Networks for Solving Nonlinear Programming Problems,"* Proceedings of IEEE ISCAS, Volume II, pp.2837-2840, 1990.

[16] Holler, M., et al., *"An Electrically Trainable Artificial Neural Network (ETANN) with 10240 Floating Gate Synapses,"* Proceedings of IEEE IJCNN, Washington DC, Volume II, pp.191-196, June 1989.

[17] Abu-Mostafa, Y. S., and Pslatis, D., *"Optical Neural Computers,"* Scientific American, 256, pp.88-95, March 1987.

[18] Farhat, N., Miyahara, S., and Lee, K. S., *"Optical Analog of Two-Dimensional Neural Networks and Their Application in Recognition of Radar Targets,"* In J. Denker (Ed.), *"AIP Conference Proceedings 151: Neural Networks for Computing,"* New York: American Institute of Physics, pp.146-152, 1986.

[19] Hecht-Nielsen, R., *"Neural-Computing: Picking the Human Brain,"* IEEE Spectrum, Volume 25, Number 3, pp.36-41, March 1988.

[20] Simpson, P., *"Artificial Neural Systems: Foundations, Paradigms, Applications, and Implementations,"* McGraw-Hill / Pergamon Press, 1990.

[21] Kung, S. Y., and J.N. Hwang, J. N., *"Parallel Architectures for Artificial Neural Networks,"* Proceedings of IEEE IJCNN, San Diego, Volume II, pp.165-172, June 1988.

[22] Akers, L. A., and Walker, M. R., *"A Limited Interconnect Synthetic Neural IC,"* Proceedings of IEEE IJCNN, San Diego, Volume II, pp.151-157, June 1988.

[23] Suzuki, Y., and Atlas, L. E., *"A Study of Regular Architectures for Digital Implementation of Neural Networks,"* Proceedings of IEEE International Symposium on Circuits and Systems, pp.82-85, May 1989.

[24] Tamberg, J., et al., *"Fully Digital Neural Network Implementation Based on Pulse Density Modulation,"* Proceedings of IEEE Custom Integrated Circuits Conference, Paper 12.7, May 1989.

[25] Weinfield, M., *"A Fully Digital Integrated CMOS Hopfield Network Including the Learning Algorithm,"* In J. G. Delgado-Farias and W. R. Moore (Editors), *"VLSI for Artificial Intelligence,"* Kluwer Academic Publishers, Boston, 1989.

[26] Yasunaga, M., et al., *"A Wafer Scale Integration Neural Network Utilizing Completely Digital Circuits,"* Proceedings of IEEE IJCNN, Washington DC, Volume II, pp.213-217, June 1989.

[27] Raffel, J., et al., *"A Generic Architecture for Wafer-Scale Neuromorphic Systems,"* Proceedings of IEEE First ICNN, San Diego, Volume III, pp.501-513, June 1987.

[28] Graf, H.P., and Jackel, L., *"Analog Electronic Neural Network Circuits,"* IEEE Circuits and Devices Magazine, Volume 5, Number 4, pp.44-49&55, July 1989.

[29] Tsividis, Y., and Anastassiou, D., *"Switched-Capacitor Neural Networks,"* Electronics Letters, Volume 23, Number 18, pp.958-959, August 1987.

[30] Rodriguez-Vazquez, A., et al., *"Switched-Capacitor Neural Networks for Linear Programming,"* Electronics Letters, Volume 24, Number 8, pp.496-498, April 1988.

[31] Horio, Y., et al., *"Speech Recognition Network with SC Neuron-Like Components,"* IEEE ISCAS, Volume I, pp.495-498, 1988.

[32] Redman-White, W., et al., *"A Limited Connectivity Switched Capacitor Analogue Neural Processing Circuit with Digital Storage of Non-Binary Input Weights,"* IEE ICANN, London, Number 313, pp.42-46, 1989.

[33] Hansen, J.E., Skelton, J.K., and Allstot, D.J., *"A Time-Multiplexed Switched-Capacitor Circuit for Neural Network Applications,"* IEEE International Symposium on Circuits and Systems, Volume II, pp.2177-2180, 1989.

[34] Vallancourt, D., and Tsividis, Y., *"Timing-Controlled Switched Analog Filters with Full Digital Programmability,"* IEEE International Symposium on Circuits and Systems, Volume II, pp.329-333, 1987.

[35] Murray, A.F., and Smith, A.V.W., *"Asynchronous VLSI Neural Networks Using Pulse-Stream Arithmetic,"* IEEE Journal of Solid-State Circuits, Volume SC-23, Number 3, pp.688-697, June 1988.

[36] Murray, A.F., et al., *"Pulse-Stream VLSI Networks Mixing Analog and Digital Techniques,"* IEEE Transactions on Neural Networks, Volume 2, Number 2, pp.193-204, March 1991.

[37] Kaehler, J.A., *"Periodic-Switched Filter Networks: A Means of Amplifying and Varying Transfer Functions,"* IEEE Journal of Solid-State Circuits, Volume SC-4, pp.225-230, August 1969.

[38] Bruton, L.T., and Pederson, R.T., *"Tunable RC-Active Filters Using Periodically Switched Conductances,"* IEEE Transactions on Circuit Theory, Volume CT-20, pp.294-301, May 1973.

[39] Liou, M.L., *"Exact Analysis of Linear Circuits Containing Periodically Operated Switches with Applications,"* IEEE Transactions on Circuit Theory, Volume CT-19, pp.146-154, March 1972.

[40] Rehan, S.E., *"Design Considerations for Switched-Resistor Active Filters,"* M.Sc. Thesis, Al-Mansoura University, Egypt, December 1987.

[41] Vallancourt, D., and Tsividis, Y., *"A Fully Programmable Sampled-Data Analog CMOS Filter with Transfer-Function Coefficients Determined by Timing,"* IEEE Journal of Solid-State Circuits, Volume SC-22, Number 6, pp.1022-1030, December 1987.

[42] Alspector, J., and Allen, R.B., *"A Neuromorphic VLSI Learning System,"* in Proceedings of the 1987 Stanford Conference, Losleben, P., (Editor), *"Advanced Research in VLSI,"* MIT Press, Cambridge, MA, pp.313-349, 1987.

[43] Tsividis, Y., and Satyanarayana, S., *"Analogue Circuits for Variable-Synapse Electronic Neural Networks,"* Electronics Letters, Volume 23, Number 24, pp.1313-1314, November 1987.

[44] Gregorian, R., and Temes, G.C., *"Analog MOS Integrated Circuits for Signal Processing,"* John-Wiley & Sons, NY, 1986.

[45] Uyemura, J.P., *"Fundamentals of MOS Digital Integrated Circuits,"* Addison-Wesley Publishing Company, 1988.

[46] Rehan, S.E., and Elmasry, M.I., *"A Novel CMOS Sampled-Data VLSI Implementation of ANNs,"* Electronics Letters, Volume 28, No. 13, pp.1216-1218, June 1992.

[47] Rumelhart, D., and McClelland, J., *"Parallel Distributed Processing: Explorations in the Microstructure of Cognition: Volumes 1 and 2,"* Cambridge: Bradford Books / MIT Press, 1986.

[48] Rehan, S.E., and Elmasry, M.I., *"VLSI Implementation of a Prototype MLP Using Novel Programmable Switched-Resistor CMOS ANN Chip,"* The 1993 World Congress on Neural Networks (WCNN'93), Portland, Oregon, pp. 96-99, July, 1993.

[49] Rehan, S.E., and Elmasry, M.I., *"Modular Switched-Resistor ANN Chip for Character Recognition Using Novel Parallel VLSI Architecture,"* Neural, Parallel & Scientific Computations journal, 1, pp.241-262, 1993.

[50] Rehan, S.E., and Elmasry, M.I., *"VLSI Implementation of Modular ANN Chip for Character Recognition,"* 36th Midwest Symposium on Circuits and Systems, Windsor, August 16-18, 1993.

[51] Rehan, S.E., and Elmasry, M.I., *"VLSI Zip-Code Recognition System Using Artificial Neural Network,"* The Fifth International Conference on Microelectronics (ICM'93), Dhahran, Saudi Arabia, pp.294-297, December 14-16, 1993.

3

A DESIGN AUTOMATION ENVIRONMENT FOR MIXED ANALOG/DIGITAL ANNS

Arun Achyuthan and Mohamed I. Elmasry

3.1 INTRODUCTION

The past decade has witnessed rapid developments in a new form of computing in the area of information processing. The developments were inspired by the architecture and operation of the biological brains, and hence they were called Artificial Neural Networks (ANNs). The computations in these systems are distinguished by their ability to learn by themselves through repetitive presentation of training data. Moreover, they are built with a large number of processing nodes which are heavily interconnected with each other and operate concurrently. These characteristics have enabled them to be excellent information processing systems and hence they have found wide use in applications such as speech processing, pattern recognition, image processing, signal processing, and robotics control.

Until recently, artificial neural systems were operated by embedding into the traditional computing environment. Even though the algorithms for computations in ANNs helped in achieving new standards in the applications that are mentioned above, their capabilities were severely constrained by the operation of the host architecture. Therefore in order to fully utilize their potentials, the ANNs need to be integrated into a hardware environment which is specifically designed to suit their architectures and operations. Such structures are implemented as application specific VLSI hardware, made up of building block circuit modules. Many implementation examples have been proposed in the past few years [1] with varying styles so as to fit the selected algorithm for a targeted application. At the top most level they can be broadly classified into digital, analog, and mixed analog/digital implementations. The ultimate choice of one style over the other is based on many different criteria such as the desired flexibility of computation, required parallelism and operational complexity, on-chip learning, and the packaging size of the implemented hardware. In this regard, digital architectures are adopted when a high degree of flexibility is desired, and the learning requires high accuracy and precision. Analog implementations are employed when the desired parallelism is high, the flexibility requirements are moderate, and the demand on accuracy and precision are low. Mixed analog/digital implementation offers a compromise between the digital and the analog approaches.

As the hardware designs mature, which is manifested by the increasing sophistication of the solutions offered at both research and commercial levels, many implementation choices become available to the designers which vary in complexity. Even though specialization helps in arriving at an optimum solution, it can lead to a situation where the whole design process, starting from the gathering of knowledge, need to be started from the very beginning when specifications and architectures are altered considerably. This can lead to higher turn-around times for new projects. The reduction in turn-around time through fast prototyping and improved search for diverse solutions is achieved by automating the design and encoding the design knowledge to higher levels of abstraction. In the digital domain, a variety of CAD tools are available for automating the designs at higher-levels[2]. Though most of them are developed for the design automation of DSP circuits and microprocessors, it is possible to adapt them for ANN applications without major effort [3]. Unfortunately, high-level synthesis of analog and mixed analog/digital systems is still at an infant state, and no tools are available that can be used for ANN applications. Non-idealities in circuit operation, the absence of clear distinction of levels in the design hierarchy, and the difficulty in multiplexing hardware modules prompt special handling and hence the designs are generally produced manually. Yet the advantages gained from reducing the time spent in hand-crafting the designs and eliminating many design cycles through correctness by construction greatly accentuates the need for automating many of the tasks in mixed analog/digital design of ANN hardware, with the overall design still conducted through user interaction.

In this chapter we describe an environment for automating the hardware design of ANNs using analog and mixed analog/digital circuit blocks. The ANN descriptions are input to the environment in the form of Data Flow Graphs. The environment provides the designer with tools for analyzing the systems for the effect of circuit non-idealities and synthesizing a high-level interconnection description by meeting the specifications on hardware complexity and operational throughput. The chapter is organized as follows. In Section 3.2 the status of hardware implementation of ANNs is briefly reviewed, with the intention

Mixed Analog/Digital Design Automation   93

of highlighting the issues involved in the selection of different choices. The design automation environment is introduced in Section 3.3, along with the automation tools that are integrated into it. The Data Flow Graph format for inputting ANN descriptions is described in Section 3.4. Section 3.5 is devoted to the description of the Analyzer tool. After justifying the need for analyzing the descriptions for the effect of circuit non-idealities, a procedure for performing quantitative analysis and deriving bounds on permissible amounts of non-idealities is explained. The section concludes with examples that demonstrate the versatility of this approach, and with details of implementing it in the environment. Section 3.6 describes the design library, which supports the design tasks by storing descriptions on circuit behavior and functions. The Synthesizer tool is described in Section 3.7, where the procedures for selecting suitable circuit blocks, partitioning the functions into analog and digital modes, and generating the hardware interconnection description by scheduling the operations and allocating circuit units, are explained. Examples of synthesizing hardware designs using this environment are provided in Section 3.8. Concluding remarks and directions for future research are provided in Section 3.9.

3.2 MIXED ANALOG/DIGITAL ANN HARDWARE

The architecture of most ANNs consist of a large number of processing nodes called *neurons*, each of which are interconnected to other neurons through weighted links called *synapses*. The neuron aggregates signals coming from the synaptic links and generates an output after *thresholding* the summed signals. Learning is achieved by repreated presentation of a set of known signals and forcing the ANN to generate a desired set of output signals, or by allowing it to form clusters by recognizing similarity in input signals. Learning is manifested in the modification of synaptic weights. The basic operations to be performed to realize these functions include multiplication, addition/subtraction, thresholding, and weight storage. Hardware implementation of ANNs proceed by designing building block circuits for performing the basic operations and interconnecting them.

3.2.1 Functional Blocks

Multiplication in analog mode is performed on voltage signals, unless one of the operands is a constant quantity. In the latter case, the variable current signal is multiplied with the resistance of a fixed resistor to generate a voltage output [15]. If both the operands are variables, multiplication is performed using differential circuit techniques. An example of such a circuit is the Gilbert multiplier [5]. The dynamic range and linearity of operation in a Gilbert multiplier are improved by using long MOS transistors at the input [6, 7]. One advantage of the Gilbert multiplier is that the product is available in the form of current, and hence it can be directly used as an operand for the addition operation, which often follows the multiplication. A Quarter Square multiplier is an example of a multiplier circuit with higher linearity and with the output signal generated in voltage form [8]. Digital multiplier blocks are operated in fixed point unless the required precision is high, in which case floating point mode is used. They may be bit-serial, serial, parallel, or pipelined. Addition is performed according to Kirchoff's laws using current signals. An operational amplifier operated in feedback mode is usually employed for providing the high impedance node required for the operation [6, 7]. Addition in analog mode is a multi-operand operation since current signals from many lines can be added together at the same node. In contrast, addition of digital data, using carry save or carry look ahead schemes, is a two operand operation. In analog mode, if the addition operation immediately follows multiplication, it is possible to merge both of the operations into an adder-multiplier block [9]. Thresholding is a non-linear operation. Most ANN algorithms require the operation to be described as a continuously differentiable function, such as the sigmoid function [8] or the hyperbolic tangent function. Although exact implementations of these functions may not be possible, close approximations using analog circuits with transistors operating in differential mode are available in both MOS and bipolar technologies [11, 21, 13, 14]. Since neurons use threshold operations immediately after performing synaptic summing, some building blocks merge these two operations into one unit [6]. In digital technology, threshold operations are implemented as look-up tables, or by piece-wise linear approximation using multipliers [3].

3.2.2 Data Storage and Update

Storage devices are required to keep synaptic weights. Moreover, in some instances many intermediate data need to be stored so that the functional block that generates them can be used in multiplexed mode. Storage of digital data using master-slave registers or RAM cells is straight forward, reliable, and easy to update. However, analog signals stored in capacitors tend to get deteriorated due to charge leakage. Even though the leakage problem cannot be solved, corrective actions can be taken through charge refreshing. Charge refreshing is achieved by storing a copy of the data in digital medium and periodically up-

Mixed Analog/Digital Design Automation 95

dating the analog values after performing Digital-to-Analog data conversions. Longer term storage of analog data without refreshing is achieved with floating gate MOS transistors that trap charges permanently at a secondary gate (floating gate) [15, 16]. The drawbacks of this storage are poor precision in updation and the limit on the maximum number of updating cycles.

3.2.3 Digital vs. Analog

Digital building blocks tend to become bigger than analog circuits, when they implement the same operation. For example, a 6 bit x 6 bit parallel multiplier requires nearly 1 sq. mm. of chip area, whereas an analog multiplier with same accuracy requires less than 0.1 sq. mm. The difference in sizes is more apparent for building blocks that implement special functions such as the threshold operation, as analog circuits make better use of the physics of device operation, and hence the device count per operation is much less. The speed of operation of functional blocks implemented in these two modes are comparable though in some instances digital operation is faster than analog operation and vice versa [10]. The size advantage makes analog technology a better candidate for ANN hardware, but they are beset with many problems. With digital circuits it is possible to achieve the desired precision and accuracy, but circuit non-idealities prevent analog operations from achieving comparable levels of accuracy. Therefore operations that require higher accuracy, such as learning operations, are usually implemented using digital circuitry. Other factors that affect the choice of the mode of operation are data storage, flexibility to share resources through multiplexing, and reliability in off-chip communication (see Table 3.1).

3.2.4 Examples of ANN Hardware Systems

So far we have examined the issues involved in the selection of analog or digital mode of operation, using examples of building block circuits. We conclude with a few implementation examples that illustrate how these issues influence the design of a hardware system.

Intel's 80170NX (ETANN) chip [16, 18] is a semi-custom design in the sense that the chip can be reconfigured to implement many different ANN systems. The chip is designed using circuits such as Gilbert multipliers, current adders, and sigmoid function generators. Weights are stored in analog form, using

Table 3.1 ANN hardware design issues and comparison of responses

Design Issue	Digital Hardware	Analog Hardware
Accuracy and precision	Can be varied to meet the requirements	Poor
Availability of building blocks	Arithmetic and logic units as universal building blocks	Many choices
Communication between processors	Reliable and flexible	Lossy; requires buffer circuits
On-chip learning	Possible	Possible only for some systems
Size of functional blocks	Large	Small
Special functions (e.g., sigmoid threshold)	Approximate realization with complex hardware	Easily realized with simple circuits
Time & resource sharing	Flexible	Not easily amenable for multiplexing
Storage and updation of synaptic weights	Accurately stored for any arbitrary length of time and easily modified	Storage is subject to degradation over time; stored data not easily modified in long term schemes

the floating gate method. The Gilbert multiplier integrates the weight storage and multiplier operation into one circuit. The inputs and outputs of the chip are in analog (voltage) form. Current addition is performed by connecting together the output nodes of the multiplier circuits. The sigmoid operation (with variable gain) is implemented using differential pairs. Switches and sample-and-hold circuits, which are provided on chip, help the synaptic operations to be performed in block-multiplexed mode: to accept input from either an external source or from the chip's output. Buffers are provided at the input for interfacing with external world. Each multiplication (and weight storage) requires only 5 transistors, making it possible to integrate more than 10,000 synapses into the chip. No learning circuitry is added to the chip and hence the weight update cycles are few. The ANNA chip [19] designed at AT&T implements a multi-layer feed-forward network [8] and uses a programmable architecture with a set of synaptic weights that are already learned. The chip is programmable in the sense that the number of neurons and the number of inputs to each of them can be varied according to the requirements. The programmable feature is made possible because all neuron activation values are made available in digital form. Weights are also stored in digital mode, but are converted to analog form for synaptic multiplication using Multiplying DACs (or MDACs, the circuits that combine D/A conversion and analog multiplication into one block). Analog computations help to achieve high parallelism (the chip is capable of performing 4096 synaptic multiplication operations in parallel) by making full use of the relatively small size of MDAC circuits. Digital form is used for programmability, reliable communication, and data storage.

In an implementation of the Deterministic Boltzmann ANN, Card and Schneider use analog circuits [20, 21] for learning and recognition. The implementation is based upon the observation that analog non-idealities can be adaptively learned, and hence the learning circuitry is designed using analog blocks. The multiplication operations in the feed-forward path, as well as in the learning circuitry, are implemented using Gilbert multiplier cells. The sigmoid operation is implemented as a multiplication operation with one of the operands held at a constant value and with saturated outputs. Weights are stored in capacitors since learning is done dynamically. Weight updating is performed in pulsed mode using small charge packets for improved precision, and the charging current is supplied by transistors operating in the sub-threshold region.

3.3 OVERVIEW OF THE DESIGN AUTOMATION ENVIRONMENT

Considering the relative merits and disadvantages of analog and digital implementation schemes, it is desirable to design ANN hardware so that the positive features are fully utilized and the deficiencies of either scheme are compensated by adopting alternative strategies. Storage reliability, computational accuracy, and the ability to share resources through multiplexing help digital systems to be used in designs where reliability and accuracy are prime, and flexibility in trading hardware utilization with operational throughput is desired to arrive at optimum circuit solutions. On the other hand, the relative compactness of analog blocks help the designer to increase concurrency of operations by using multiple instances of the blocks within the constraints of silicon real-estate utilization. But circuit non-idealities limit their application in accurate computations or when resource sharing is desired. Therefore the main challenge in the design of ANN hardware is in finding an ideal mixture of digital and analog blocks, so that optimum resource utilization is achieved with increased computational speed, by satisfying the constraints on accuracy and precision. It is expected that design automation methods help in meeting this challenge by synthesizing a high-level interconnection description of building block circuits with minimum turn-around time.

Many heuristic techniques already exist for synthesis of digital systems [2], and SPAID-N [3] is a program for ANN applications. But the efficiency of the architectures generated by SPAID-N is limited by the size of processing units and hence SPAID-N cannot improve performance throughput of the synthesized hardware by increasing the concurrency of operations. The corresponding scene is even worse in analog domain because only a few tools are available for high-level synthesis [22, 23] and none of them have applied design of ANN systems. Therefore there is a need to develop a new methodology for mixed analog/digital synthesis which addresses the requirements of ANN applications. The important requirements for design automation of such systems are:

- A procedure for the selection of analog circuit blocks which meet the performance constraints imposed by the ANN system. This procedure should take care of the non-idealities inherent in the operation of analog circuits, and select only those circuits whose non-idealities are within acceptable limits. The procedure should be such that it is applicable to most ANN systems quickly and with minimum modification.

- A method for partitioning the operations to analog and digital modes, such that the overheads in mixing them are minimum. The overheads are mainly due to circuits that convert data between the two modes. Therefore the reduction is achieved by minimizing the interfaces between the partitioned sections.

- Adoption of different strategies for scheduling and allocation of digital and analog sections. The conflicting requirements for design automation of the two styles are therefore separately addressed, but at the same time the design goals are still met with a common approach for integrating the designed sections.

- The synthesized interconnection structure should preserve the architectural regularity in ANN systems. They should also make use of building block hardware.

Mixed Analog/Digital Design Automation

- The synthesis procedure should be flexible so that different solutions can be explored with varying requirements on operational parallelism and hardware size. Though it is possible to introduce mechanisms to shift emphasis from one requirement to another, this implies the need for user intervention for guiding the search for a better solution.
- Many ANN systems are inherently asynchronous in operation. In order to make use of this mode, the synthesis procedure should have the facility to incorporate asynchronously operating data paths into the synthesized hardware.

The methodology is integrated into an environment which supports a number of automation tools for performing various tasks which are mentioned above. The environment also provides facilities for iterative execution of these tasks to arrive at an optimum design.

A schematic overview of the design automation environment for hardware implementation of ANNs is given in Figure 3.1 The main components of the environment are: the Data Flow Graph (DFG), the Analyzer, the Hardware Synthesizer, and the Design Library. An internally maintained database links the activities of various components. The specification of the ANN is entered in the form of a DFG, the details of which are provided in the following section. User interfaces to the Analyzer and the Synthesizer are provided for guiding the hardware mapping procedure.

Analyzer

The Analyzer performs quantitative evaluation of ANNs and generates specifications on allowable amounts of non-idealities. Taylor-series approximation methods are used for the analysis. The errors due to non-idealities are modeled as random quantities. The Analyzer divides the original DFG input description into compact sub-graphs, and the Taylor-series expressions are extracted symbolically. The maximum limits on errors are determined using optimization methods. The error limits returned by the optimizer are back-annotated to specifications on building block circuits.

Synthesizer

The Synthesizer screens and selects building block circuits, reduces the number of analog-digital interfaces, schedules the sequence of operations, and binds the DFG nodes to circuit blocks. The final hardware description is generated by the Synthesizer as an interconnection of these circuit blocks. The other functions

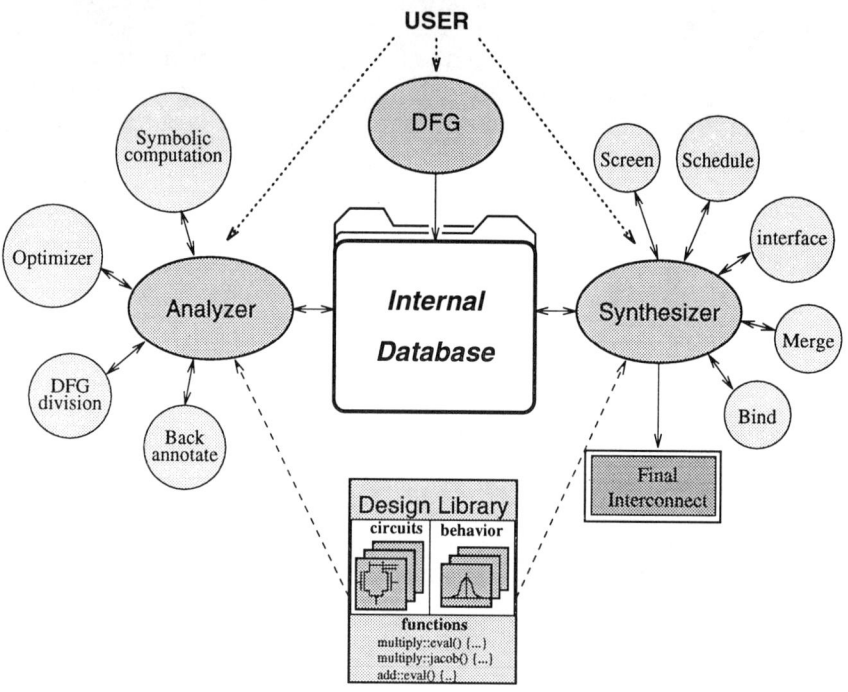

Figure 3.1 The Design Automation Environment

of the Synthesizer include merging multiple functions to one circuit and inserting additional components to interface analog circuits that have conflicting modes and ranges of operations.

Design Library

The Design Library stores information on operators and circuits, and is consulted by the Analyzer and the Synthesizer during hardware mapping. The information on operators contain descriptions for computing the input-output relations desired from them, and for computing the partial differentials of the operations. The circuit blocks are stored in the Library categorized by the functions they perform. Behavioral models on non-idealities from each of these circuit blocks are also stored in the Design Library.

3.4 DATA FLOW GRAPH

The ANN is described to the design environment in the form of a Data Flow Graph (DFG). A DFG is a directed graph $G(N,E)$ where N is the set of nodes and E is the set of edges. Each edge belonging to the set E is directed and is described as a pair (s,d), where s is the *source node* of the edge and d is its *destination* node. Both s and d belong to the set N. Each edge is directed, and indicates the flow of data from s to d. Every node in N belongs to either of the three categories: O, D, and V. The subset of nodes belonging to category O represent *operators*, such as multiplication, addition, etc. Each node of this category may have multiple edges incident on it. The nodes belonging to D represent storage operation. The stored data are either provided to the ANN externally, or are computed during the system operation. A node that receives data from outside of the system normally has no edge incident upon it. The other type of D nodes have only one incoming edge. Each of the O and D nodes has only one outgoing edge. The third category of nodes (type V) are called virtual nodes. Each node of this category has one incoming edge and two or more outgoing edges, with the data content in each of the outgoing edges being the same. A V node is inserted in the graph when the data from an O or a D node has to be branched out into multiple paths.

A DFG represents the complete set of operations performed by the ANN system. Therefore it is assumed that the ANN architecture, in terms of the number of neurons, synapses, and their interconnections, is finalized prior to the hardware implementation step. Any hierarchy in the algorithmic description of the system is to be flattened before describing it in the DFG format. An example of a DFG implementing a hierarchy of equations is shown in Figure 3.2.

For every node of category O, the Design Library stores a function for emulating its operation by computing on its operands. For every D node, the statistical information on input and stored data is to be supplied by the user along with its description in the DFG.

The DFG description is different from a Signal Flow Graph [24] format that is used in the architectural synthesis of digital systems. A restriction in the latter is that the graph should not have any directed cycles without a storage node in it. In a DFG describing ANN systems, such cycles are deliberately introduced to indicate that the execution of operations along the path forming the cycle is in asynchronous mode.

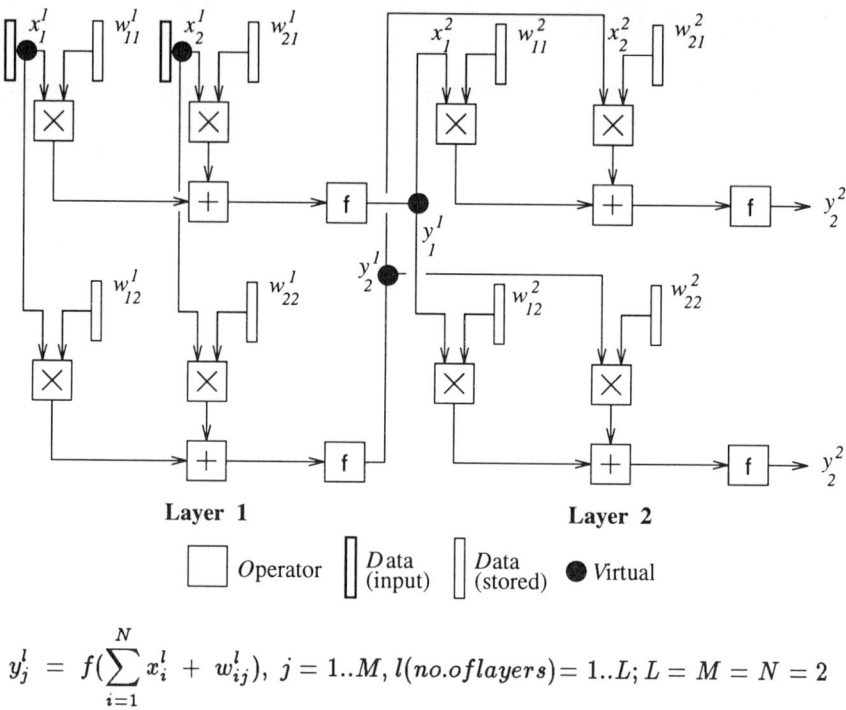

Figure 3.2 Example of a Data Flow Graph for an ANN equation

$$y_j^l = f(\sum_{i=1}^{N} x_i^l + w_{ij}^l), \; j = 1..M, \; l(no. of layers) = 1..L; \; L = M = N = 2$$

3.5 THE ANALYZER

Most of the ANN systems learn the synaptic weights by adapting to the patterns in the input vector and to the environment in which the computations are performed. In that sense it is possible that many of them are capable of tolerating the imperfections in computation by adapting to those situations. On the other hand it is also known empirically that many learning paradigms, such as the Back-Propagation algorithm, require high accuracy and precision to adaptively modify the synaptic weights. In hardware implementations computational imperfections arise due to non-idealities that are inherent in the operation of building block circuits. Non-idealities are essentially undesired deviations from perfection, and hence they can be viewed as *errors* in computation. Therefore studying the effects of non-idealities is equivalent to determining the amount of error that can be tolerated from each operation. Since each ANN behaves differently to the non-ideal effects, it is therefore necessary to study

Mixed Analog/Digital Design Automation 103

the performance of each of them with quantified models of these effects. Such an evaluation helps in determining the suitability of using a circuit block for implementing an ANN operation by generating maximum limits on the non-idealities. In our design automation environment these tasks are performed by the *Analyzer* tool.

3.5.1 Analysis of ANNs for Effect of Non-idealities

The non-ideal effects can be analyzed by simulating the operation of the ANN with hardware models replacing the operations such as multiplication, addition, etc. [25, 26, 27]. But computer simulations only help in determining pass or fail for the given hardware model. They are also very time consuming. Therefore it is not judicious to use simulation method for automated analysis, even though they can still be employed for verification of limits generated by other means. Information theory [28] and related techniques can also be used for determining the effect of non-ideal behavior. The mutual information between the desired response from an operator and the actual response corrupted by noise (due to non-ideality errors) can be used to model the information content in system response, and hence to evaluate the effect of noise or to set limits on the variance of noise distribution for satisfactory system performance. Though this method has been applied for studying quantization effects in digital implementations of ANNs [29], the calculation of mutual information at various processing stages of the ANN tend to be cumbersome as the number of noise sources increase. Moreover, it is difficult to determine how the effects of noise (error) from one circuit propagates to the other parts of the ANN system, and hence to set limits on each error based upon the propagated effects.

We use the Taylor-series approximation method for error analysis. Using this technique the error terms can be separated from ideal operations, and the error accumulated from a series of operations is calculated in terms of the error originating from each of these operations and propagated to later stages.

3.5.1.1 Quantitative Error Analysis

The Taylor-series method has been used in the past not only as a method for calculating moments in tolerance analysis [30], but also in the analysis of digital systems for error due to finite computational precision [31, 32]. The error terms are derived by applying the method to the equations describing the operation of ANN systems. In general, such an equation is described as:

$$y = \Phi^{(n)}(X^{(m)}), \quad \text{where} \quad \Phi^{(n)} \equiv \{\phi_1, \phi_2, \ldots, \phi_i, \ldots, \phi_n\}$$
$$\text{and} \quad X^{(m)} \equiv \{x_1, x_2, \ldots, x_j, \ldots, x_m\} \quad (3.1)$$

In this equation there are n arithmetic operators (ϕ_i's) operating upon m variables (the x_j's) in succession. The variable on the left hand side, y, is the result from applying the compound operator Φ to the set of inputs X. If each operator generates an error e_{ϕ_i} and if an amount of error e_{x_j} is associated with each variable, the equation (3.1) can be modified as:

$$\tilde{y} = \phi_n(\phi_{n-1}(\Phi^{(n-2)}(x_1 + \epsilon_{x_1}, X^{(m-1)} + \epsilon_X{}^{(m-1)}) + \epsilon_\Phi{}^{(n-2)}) + \epsilon_{\phi_{n-1}}) + \epsilon_{\phi_n} \quad (3.2)$$

where $\tilde{y}$ is the new erroneous output. Equation (3.2) is expanded using first order Taylor approximation, for separating the ideal expressions from the error terms. We get

$$\tilde{y} \approx \Phi(X) + \Phi'_n(X)\epsilon_{\phi_n} + \Phi'_{n-1}(X)\epsilon_{\phi_{n-1}} + \cdots + \Phi'_m(X)\epsilon_{x_m}$$
$$+ \Phi'_{m-1}(X)\epsilon_{x_{m-1}} + \cdots \quad (3.3)$$

The superscripts are omitted in the above equation for clarity and simplicity. The first term is the expression for the output under ideal conditions. The sum of the contributions from the remaining terms is the accumulated error on output y. Each term there is obtained by partial differentiation of the ideal expression with respect to the corresponding erroneous quantity. Therefore the error terms of (3.3) can be written concisely as [32]:

$$\epsilon_y \approx \sum_{j=1}^{m} \epsilon_{x_j} \frac{\partial y}{\partial x_j} + \sum_{i=1}^{n} \epsilon_{\phi_i} \frac{\partial y}{\partial \phi_i} \quad (3.4)$$

where ϵ_y is the accumulated error at y. The derivation of the expression for accumulated error is carried out based upon the assumption that the range of errors is small compared to the nominal dynamic range of the operators, and the range of values that are assigned to the variables. These assumptions are justified in a hardware domain as the errors are caused by mild deviations from ideal operations that are desired from building block circuits. Since the errors are small, the Taylor-series expansion can be limited to first order, as the contributions from higher order terms will be negligible.

In order to quantify the accumulated error, the individual error variables need to be modeled. For digital circuits the error is due to truncating (or rounding) the data by throwing away lower significant bits. In analog implementations the non-idealities arise from random variations in the fabrication process. Therefore it is reasonable to assume that the errors are random variables and that they are fairly uncorrelated [33]. With this assumption (some details on modeling the non-idealities are explained in the next section) the mean and variance of the distribution of each ϵ_{ϕ_i} and ϵ_{x_j} is substituted in (3.4) and the mean square value of ϵ_y can be calculated.

Once an expression for the compound error is obtained, the next step is to determine the maximum errors from the operators and data so that the mean square value of accumulated error is within acceptable limits. Graphical methods are used in [32] to plot the mean square error in terms of the number of bits of precision. In analog implementations there are many error terms contributing to the expression for the compound error, and hence plotting methods are cumbersome. A more feasible method for determining the worst case error combination is to formulate this task as a constrained optimization problem. The optimization method can not only handle multiple numbers of variables, it also provides a systematic procedure to arrive at a good solution point. Since the aim is to determine the maximum values for each error, the objective function for the problem is formed as a weighted sum of the error variables. The mean and variance of error distribution are used as the optimization variables. The worst case combination is obtained by maximizing the value of this objective function. The constraint equations for the optimization are of two types. The equations of the first type are obtained from the Taylor-series expressions for calculating the compound errors. These are equality constraints and serve the purpose of connecting compound errors to basic error variables without applying back substitution. The other constraints set limits on the mean square values of compound errors which are critical to the operation of the ANN, and hence are written as inequalities.

The optimization problem is stated as following:

$$Maximize \quad \sum_i \left(A_i v_{\epsilon_{\phi_i}} + B_i |\mu_{\epsilon_{\phi_i}}| \right) + \sum_j \left(C_j v_{\epsilon_{x_j}} + B_j |\mu_{\epsilon_{x_j}}| \right)$$

subject to

$$\begin{aligned} c_j(v, \mu) &= 0, \\ c_k(v, \mu) &\geq 0 \end{aligned} \quad (3.5)$$

where

$$v \equiv \{v_{\epsilon_{\phi_1}}, \ldots, v_{\epsilon_{\phi_i}}, \ldots, v_{\epsilon_{\phi_n}}, v_{\epsilon_{x_1}}, \ldots, v_{\epsilon_{x_j}}, \ldots v_{\epsilon_{x_m}}\}$$

and
$$\mu \equiv \{\mu_{\epsilon_{\phi_1}}, \ldots, \mu_{\epsilon_{\phi_i}}, \ldots, \mu_{\epsilon_{\phi_n}}, \mu_{\epsilon_{x_1}}, \ldots, \mu_{\epsilon_{x_j}}, \ldots \mu_{\epsilon_{x_m}}\}$$

Here, c_j is the set of equality constraints, formed out of Taylor-series equations and c_k is the set of performance driven inequality constraints provided by the user. Though the objective function in (3.5) is linear, the constraint equations are non-linear, because non-linear computations may occur in the calculation of partial derivatives for Taylor-series evaluation. Therefore the optimization problem represented by (3.5) is non-linear.

A_i, B_i, C_j, and D_j are constants, used for weighting the error variables. The weights are assigned so as to selectively emphasize the effect of certain variables. Typically, large weights are assigned to variables which are suspected to be critical. Since non-linear optimization problems are generally solved using gradient search methods, large weights provide large gradients, and hence the critical variable values get pushed closer to their limits.

To summarize, we observe that the following tasks need to be performed for quantitative analysis:

- Modeling and derivation of the basic errors ϵ_{ϕ_i} and ϵ_{x_j}, considering them as random quantities.
- Computation of the partial derivatives, and application of statistical methods to calculate the mean and variance of the derivative expressions [34].
- Identification of compound errors which critically affect the performance of the ANN, and formulation of constraints which restrict their values to remain within acceptable limits.
- Computation of a worst case combination of values for the errors ϵ_{ϕ_i} and ϵ_{x_j} that satisfy the constraints on the compound errors.

3.5.1.2 Application to ANN Systems

The quantitative error analysis method described above is a generalized technique for the evaluation of various ANN systems. When this method is applied to various ANN paradigms, only the performance dependent constraints need to be specified differently. The procedure for computation of compound errors is the same, irrespective of the architecture of the ANN or the algorithm that is used. The application of the analysis procedure is illustrated in this section with the help of three different ANN systems. They are:

Mixed Analog/Digital Design Automation

- The Back-Propagation ANNs [8]
- Hopfield Networks [35]
- Deterministic Boltzmann ANNs employing Mean Field Theory for learning [20]

Back-Propagation Algorithm

The equations describing the Back-Propagation algorithm [8] and the corresponding error equations are given below:

Forward Propagation

$$x_j^{l+1} = f\left(\sum_{i=1}^{N} w_{ij}^{l+1} \times x_i^l + \theta_j^l\right), \quad j = 1...M, \; l = 1...L \quad (3.6)$$

$$\epsilon_{x_j^{l+1}} = \sum_{i=1}^{N} \epsilon_{x_i^l} \dot{f}_i^{l+1} w_{ij}^{l+1} + \sum_{i=1}^{N} \epsilon_{w_{ij}^{l+1}} \dot{f}_i^{l+1} x_i^l + \sum_{i=1}^{N} \epsilon_{\times_i} \dot{f}_i^{l+1} +$$
$$\sum_{i=1}^{N} \epsilon_{+_i} \dot{f}_i^{l+1} + \epsilon_{\theta_j^{l+1}} \dot{f}_i^{l+1} + \epsilon_{f(\cdot)} \quad (3.7)$$

Back-Propagated Error (at output layer)

$$\delta_j^L = \dot{f}_j^L \times (t_j - x_j^L) \quad (3.8)$$

$$\epsilon_{\delta_j^L} = \epsilon_{x_j^L}((t_j - x_j^L)\ddot{f}_j^L - \dot{f}_j^L) + \epsilon_{t_j}\dot{f}_j^L + \epsilon_{-_j}\dot{f}_j^L + \epsilon_{f_j^L}(t_j - x_j^L)$$
$$+ \epsilon_{\times} \quad (3.9)$$

Back-Propagated Error (at hidden layers)

$$\delta_i^l = \dot{f}_i^l \times \sum_{j=1}^{M} \delta_j^{l+1} \times w_{ij}^{l+1} \quad (3.10)$$

$$\epsilon_{\delta_i^l} = \sum_{j=1}^{M} \epsilon_{\delta_j^{l+1}} \dot{f}_i^l w_{ij}^{l+1} + \sum_{j=1}^{M} \epsilon_{w_{ij}^{l+1}} \dot{f}_i^l \delta_j^{l+1} + \sum_{j=1}^{M} \epsilon_{\times_j} \dot{f}_i^l + \sum_{i=1}^{M-1} \epsilon_{+_j} \dot{f}_i^l +$$
$$\epsilon_{x_i^l} \sum_{j=1}^{M} \ddot{f}_i^l \delta_j^{l+1} w_{ij}^{l+1} + \epsilon_{f_i^l} \sum_{j=1}^{M} \delta_j^{l+1} w_{ij}^{l+1} + \epsilon_{\times} \quad (3.11)$$

Weight Updating

$$\Delta w_{ij}^l = \eta \times \delta_j^l \times x_i^{l-1}, \quad (3.12)$$

$$\epsilon_{\Delta w_{ij}^l} = \epsilon_{x_i^{l-1}} \eta \delta_j^l + \epsilon_{\delta_j^l} \eta x_i^{l-1} + \epsilon_{\times_1} \eta + \epsilon_{\times_2} \quad (3.13)$$

In Equations (2.6)-(3.13) l is the layer ($l = 1$ denotes input layer and $l = L$ denotes output layer), x_i^l is the signal from i^{th} neuron at the lth layer, w_{ij}^{l+1} is the synaptic weight between i^{th} neuron of lth layer and j^{th} neuron of $(l+1)^{th}$ layer and θ_j^{l+1} is the bias for j^{th} neuron of $(l+1)^{th}$ layer. $f(\cdot)$ is a non-linear threshold function and f_i^l is its derivative at i^{th} neuron with respect to the synaptically summed input to that neuron. t_j is the target signal (the signal to be learned) for the jth neuron on output layer, δ_i is the back-propagated error at i^{th} neuron as a result of the difference between target signal and activated signal at the output neurons. Δw_{ij} is the weight update quantity, and η is the rate of updating. $\epsilon_\times$ is the multiplication error, ϵ_+ is the error in addition, and ϵ_f is the error in threshold operation.

Since learning is most critically affected by the error in weight updating, and since gradient descent for learning is dependent upon the estimate of all weight update values, the constraint is applied to a ratio of mean square value of $\epsilon_{\Delta w}$ and mean square of all weight changes. The constraint is therefore written as [32]:

$$c \geq \frac{E[\epsilon_{\Delta w}^2]}{E[\Delta w^2]} \quad (3.14)$$

where c is a constant whose value is set to much less than 1.0.

Hopfield Networks

A Hopfield network is a single layer ANN in which all the neurons are connected to each other, except to themselves. The dynamic behavior of the network with n neurons is described as

$$C_i \dot{u}_i = -\frac{u_i}{R_i} + \sum_{j=1}^{n} T_{ij} V_j + I_i, \quad i = 1...n \quad (3.15)$$

where V_i is the neuron activity, u_i is the input to a neuron, I_i is the external input, C_i and R_i are the damping capacitance and resistance, respectively, and T_{ij} is the synaptic connection strength. Usually the Hopfield network is operated with a set of synaptic connection strengths which are already learned.

In his study of the equilibria of Hopfield networks, Vidyasagar [36] reports that asymptotically stable equilibria are found only at the corners (described as a vector of neuron activations V_i) of an n-dimensional hypercube $H = (0,1)^n$ (where n is the number of neurons), under the assumption that the neuron transfer function $V_i = g(\lambda u_i)$ is a switching function with $\lambda \to \infty$. Also, under stable conditions, $\dot{u}_i = 0$. He further states that, with unity values assigned to the capacitances and resistances, the equilibrium approaches a corner V only if V satisfies the parity condition defined as: Let $u = TV + I$. Then

$$u_i > 0 \ if \ V_i = 1, \ and \ u_i < 0 \ if \ V_i = 0. \tag{3.16}$$

where u is the neuron input vector, T is the synaptic connection matrix, V is the neuron activation vector, and I is the external input vector.

This condition for equilibria is used to form the performance constraint for the Hopfield network, when the network is implemented with non-ideal hardware. The equation for aggregate error, obtained from the Taylor-series approximation of Equation (3.15) with $\dot{u}_i = 0$ is:

$$\epsilon_{u_i} = \sum_{j=1}^{n} \epsilon_{T_{ij}} V_j R_i + \sum_{j=1}^{n} \epsilon_{V_j} T_{ij} R_i + \sum_{j=1}^{n} \epsilon_{\times_j} R_i + \sum_{j=1}^{n-1} \epsilon_{+_j} R_i +$$

$$\epsilon_{I_i} R_i + \epsilon_+ R_i + \epsilon_{R_i}(\sum_{j=1}^{n} T_{ij} V_j + I_i) + \epsilon_\times \tag{3.17}$$

If the aggregate error at the input of any neuron is high, the parity condition stated in (3.16) will be violated. This happens if the error ϵ_{u_i} that is added to u_i reverses its polarity, and hence the desired parity condition for certain vertices will never be satisfied. The inequality constraint for evaluation of Hopfield ANNs should therefore relate the mean square error of neuron inputs to the mean square value of the inputs, and is stated as:

$$E[u_i^2] - E[\epsilon_{u_i}^2] \geq 0 \tag{3.18}$$

Deterministic Boltzmann ANNs

Deterministic Boltzmann ANNs are multi-layer networks, but use Mean Field Theory or Contrastive Hebbian Rule [21] for learning of synaptic weights. Unlike the Back-Propagation algorithm, for which only neurons between adjacent layers are allowed to be synaptically connected, there is no restriction on connections between neurons in a Deterministic Boltzmann ANN (except that the weights between pairs of neurons are symmetric and that no neuron is connected to itself). Learning is supervised and is carried out in two phases. In the first phase, the activation signals from all the neurons in the input and the output layers are *clamped* to the input vector and the target vector, respectively, and the activation signals from the remaining neurons are allowed to settle to a stable value. In the second phase, the clamping on output layer neurons is removed and the activations from all the neurons, except those in the input layer, are allowed to settle to a stable value. The activations from the input neurons are still clamped to the input vector at this phase. At either phase, the neuron activations are calculated using a relation similar to (3.6) described under the Back-Propagation algorithm. The relation is:

$$V_j = f\left(\sum_j w_{ij} \times V_j\right). \qquad (3.19)$$

Layer designations are omitted in the equation since they are irrelevant. The weight of the synapse connecting neuron i to neuron j is updated using Hebbian learning by a quantity which is the difference between the product of activations from these neurons in the clamped and unclamped phases. The weight update equation is:

$$\Delta w_{ij} = \eta \times (V_i^+ \times V_j^+ - V_i^- \times V_j^-) \qquad (3.20)$$

where the superscript '+' denotes clamped phase and '-' denotes unclamped phase, and η is a learning constant. Learning is said to have converged when the difference between clamped and unclamped activations is reduced and the value of Δw_{ij} from successive iterations falls below a predefined constant. Schneider and Card note in [21] that the matching between clamped and unclamped phases should be good, so that when the network has learned a pattern perfectly the weight change is reduced to zero. Mismatch between the phases arises as a result of variations in the neuron activation values, and hence the condition for good match implies that the propagated error on activation values ($V's$) should be very small. The effect of mismatch also manifests if the error in calculating the difference between the clamped and the unclamped activations is larger than the predefined constant for convergence. The above two conditions are encapsulated in the following two constraints:

$$E[\epsilon_V^2] \leq k_1, \text{ and } E[\epsilon_{\Delta w}^2] \leq k_2 \qquad (3.21)$$

where k_1 is the required accuracy on neuron activations, and k_2 is the predefined convergence constant. The expression for ϵ_V is the same as that of ϵ_x and is

Mixed Analog/Digital Design Automation 111

described by (3.7). The Taylor-series equation for $e_{\Delta w_{ij}}$ is:

$$\epsilon_{\Delta w_{ij}} = \epsilon_{V_i^+} \eta V_j^+ + \epsilon_{V_j^+} \eta V_i^+ + \epsilon_{\times_+} \eta + \epsilon_{V_i^-} \eta V_j^- + \epsilon_{V_j^-} \eta V_i^- + \epsilon_{\times_-} \eta + \epsilon_- \eta + \epsilon_\times \tag{3.22}$$

3.5.2 Implementation of the Analyzer

So far we have described the theory behind the analysis procedure and how it can be applied to many different ANNs. In this section we explain the techniques for automating the procedure.

The DFG is input to the environment in textual form. This description has three sections. The first section defines the nodes (vertices) and the parameters associated with each of them. The second section is for defining connectivity between nodes. The constraint equations are described in the third section. The internal database converts the textual description into a graphical form. Before the conversion is made, the composite description is divided into many sub-graphs. The division is performed so that each sub-graph is compact enough for performing the analysis. As a result of this division each sub-graph roughly corresponds to an ANN equation. The sub-graphs are constructed so that none of them contains any re-convergent fan-out paths, and hence they are stored as tree structures, with the last node in the path forming the root, and the data entry nodes forming the leaves.

The computations of partial derivatives, using the statistical values of error variables, are performed by traversing the trees in bottom-up mode. The computation is symbolic, in the sense that only the variable expressions are generated at each node, and numerical results are calculated only when they are required. At each node the procedures stored in the design library and appropriate to the operation performed at that node are applied recursively to the child nodes and the partial results are accumulated. The expression for the compound error is computed in the same fashion at the root node, and the variables are stored in a list. The inequality constraints, specified as textual equation in the input file, are also converted to calculation trees before storing them in the internal database. The optimization is performed using the routines from NAG Fortran library [37]. They solve the optimization problem iteratively, using a modified gradient search procedure. The intermediate values of the objective function and the constraints, which are required at each iteration, are calculated by visiting the root nodes of sub-graphs and using the variables stored in the

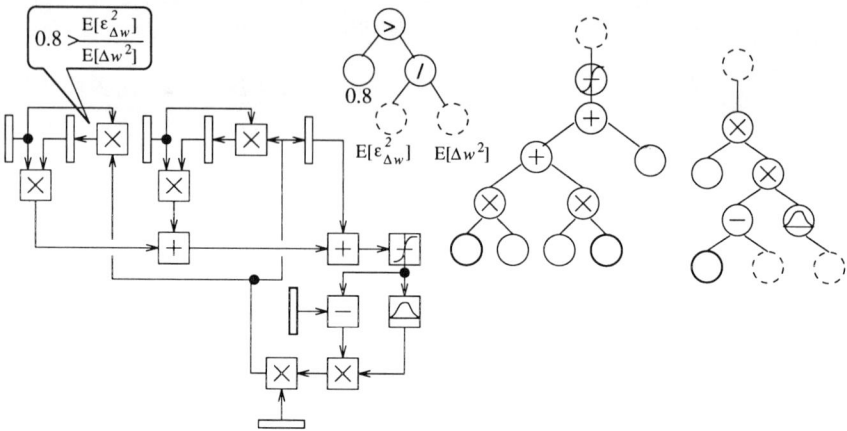

Figure 3.3 Division of DFG and formation of sub-graphs

list. The formation of sub-graphs and calculation trees are illustrated in Figure 3.3.

3.6 THE DESIGN LIBRARY

The design library supports the Analyzer and the Synthesizer tools by storing information on operators and building block circuits. At the functional level, descriptions are stored for ANN operations for computing the output from input data. These descriptions are used while examining a tree node for generating Taylor-series expressions and the computations are applied to the child nodes [38].

Building block descriptions contain information on the type of primitive operation performed by the circuit, the operational delay, and the hardware cost. If the circuit is multi-functional, all the operations performed by it are listed, in the order by which they are executed. The hardware cost is used by the Synthesizer for calculating the cost of synthesis. It is a relative figure and is specified as an integer value. For digital building blocks, the cost is specified as a function of the required resolution in number of bits. For analog blocks, the type of signal at the input and output are also stored along with their dynamic ranges. If the building block is capable of performing the operation with multiple number of operands, the maximum number of operands that it can accommodate is also specified.

Behavioral models describe the non-idealities from circuit operation. These models are generated using circuit simulation and curve fitting. Even though the performance of a circuit is dependent upon the circuit and device parameters, the results are indicative of the typical non-ideal effects expected from its operation. Monte-Carlo methods are used for simulations to take care of random deviations in device parameters, which are in turn caused by variations in fabrication process. The results from a set of simulations are fitted on a curve that is expected from a circuit operating under ideal conditions. A plot of the distribution of error from fitting is used as the model for the circuit's non-ideal behavior. An example of such a plot from the modeling of a Gilbert multiplier is shown in Figure 3.4. It is obtained from plotting output current and is indica-

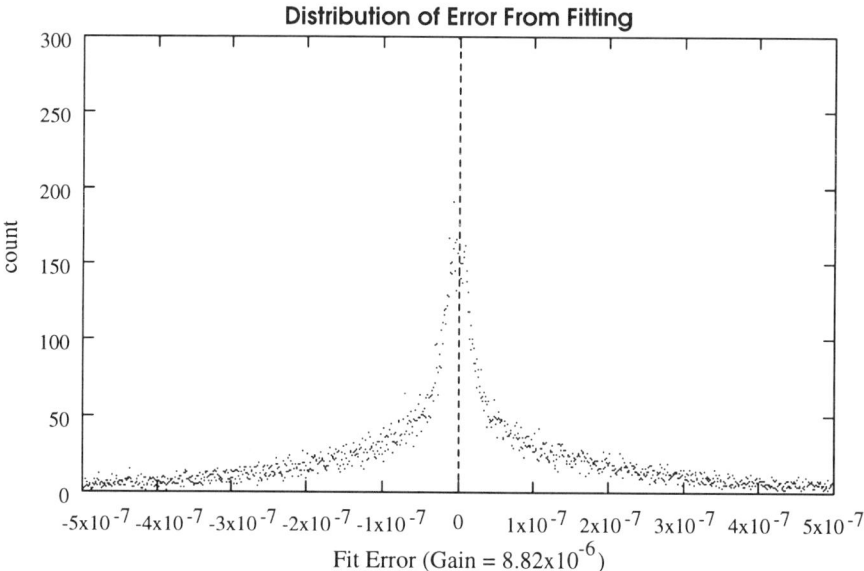

Figure 3.4 Distribution plot of errors from fitting simulated values of output current ($Isubout$) for the Gilbert Multiplier

tive of the errors in transconductance gain. The simulations were conducted by uniformly varying both the inputs to the multiplier between -1 volt and +1 volt.

In order to model error due to charge leakage from a capacitor, the following decay equation is used.

$$\epsilon_{ch} \approx v(1 - e^{-t/T}) \qquad (3.23)$$

where ϵ_{ch} is the error due to decay, v is the voltage stored in the capacitor, and t is the time elapsed since last storage or refresh. T is the decay time-constant, which is the non-ideal parameter in this equation. Assuming that a bipolar charge storage scheme is used, and that the voltage stored in the capacitor is uniformly distributed (taking the entire period of the ANN operation into consideration), the distribution of decay error can be approximated to be random, if the stored voltages are sampled or updated at random intervals. Equation (3.23) is used under these assumptions to determine the statistical distribution of ϵ_{ch}, in terms of the distribution statistics on v and t, for different values of T.

3.7 THE SYNTHESIZER

The goal of synthesis is to generate a description of interconnected building block hardware from a high level specification of a given system, and meeting a set of constraints on hardware size and performance throughput. The error limits on operations, which are generated by the Analyzer, are used as specifications for selecting suitable building blocks from the design library. Since digital blocks are capable of meeting the required accuracy and precision, these specifications are only applied to analog circuits. If no analog block is capable of implementing an operation, a digital block is used there. Since the building blocks can be interconnected in many different ways, the task of the synthesizer is to evaluate each combination, reject the ones which fall below the expectations, and select a configuration that meets the performance constraints with minimum hardware utilization.

The synthesis program is intended to be operated in such a way that it strives to preserve the parallel and distributed computing capabilities of ANNs using modular structures. But preservation of parallelism implies that many operations are done concurrently using multiple instances of hardware units. In order for the computing to be distributed, the sharing of storage units should be minimum, and each of them should be closely associated with only one processing unit. Modularity can be achieved by imposing regularity in interconnection. The preservation of all these features exactly as it is would require a large amount of hardware resources to be used. In practical VLSI designs the silicon area for implementing the circuits is constrained, and hence a compromise needs to be reached between the required performance and attainable complexity. One way of achieving the middle ground is to *reduce* the concurrency of operations and let the operations *share* the hardware blocks. Although reduction of concurrency results in degradation in throughput from the implemented system, the relation between concurrency and hardware resource utilization is not linear due to two reasons. Firstly, the execution of operations follow a certain order as some operations are dependent upon the data generated from earlier stages, and hence their execution need to be deferred until the required data is generated and is ready to be used. Therefore there is no speed advantage from allocating separate hardware blocks to perform each operation. The other reason for non-linear dependency of hardware utilization and concurrency is that many sets of operations can be executed using *functional pipelining*. Pipelining is helpful in executing *blocks* of operations at successive time steps by resource sharing and hence the increase in execution time due to sequentialization is minimum. This situation is illustrated in Figure 3.5. Here the increase in the duration of an execution cycle due to sequentialization of the multiplication and the addition operations is minimum because operations such as those at nodes 3, 6, and 8 are pipelined with preceding operations at nodes 2, 5, and 7.

Synthesizing hardware interconnection descriptions is equivalent to searching a design space to find a good solution. In mixed analog/digital implementations, the design space to be searched is multi-dimensional. We have already seen one dimension, namely trading concurrency with hardware resource utilization. The other dimensions of the design space are:

- The circuits that are suitable for hardware implementation of a given ANN system. The boundaries of this dimension[1] are set by the specifications on the amount of non-idealities that could be tolerated.

- The mode of operation of ANN functions. Since the operations can be either analog or digital, a number of schemes with different mixtures of hardware blocks operating in these two modes are available from this dimension.

The design space is searched following a set of directives from the designer that specify the size of synthesized hardware and the required throughput from it. If the learning is not performed in real time, there is no specified limit on

---
[1] For the given ANN and the associated specifications. The boundary could be different for a different set of specifications.

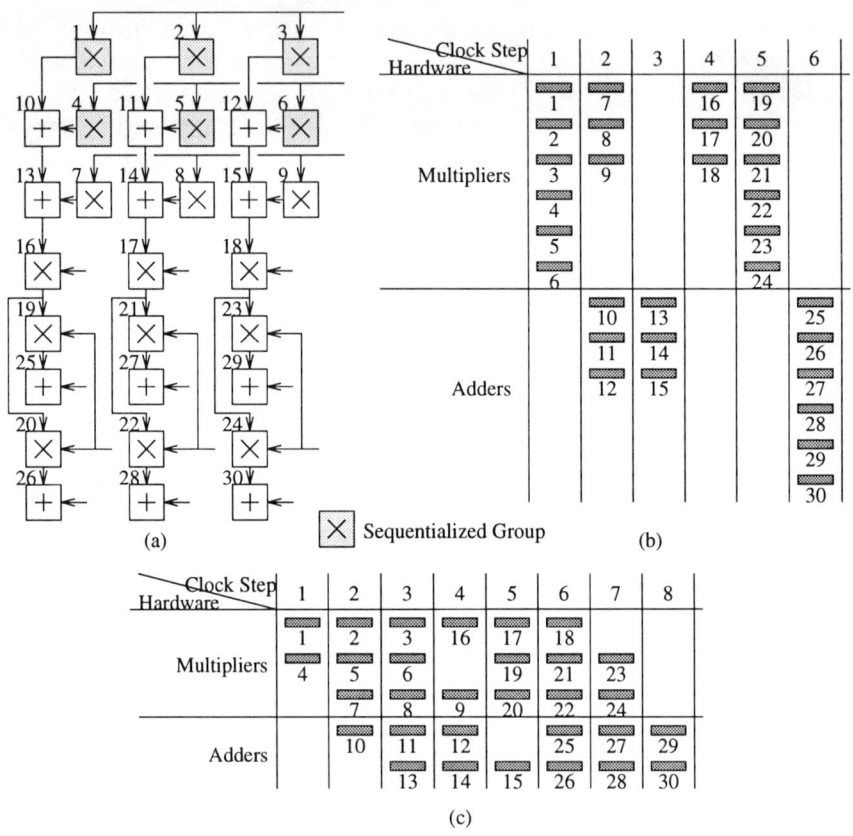

Figure 3.5 Selective sequentialization of ANN operations (a) DFG (b) schedule for parallel implementation (c) pipeline schedule with multiplications 1-6 sequentialized

learning time, but each iteration should be completed in a reasonably small time period. For on-line learning and recognition, the synthesized hardware should complete the operations within the specified time T. In any case the objective is that a joint cost function C, composed of hardware costs and execution time, be minimized during synthesis.

The procedures for searching the design space are fitted into a general framework of synthesizing digital systems, due to the relative maturity of the techniques that are applied. But the design strategy accommodates the requirements for interconnecting analog blocks. One such requirement is the limit on the storage time for analog data. Due to the problems arising from charge

Mixed Analog/Digital Design Automation

leakage in capacitors, the stored data cannot be held for a long time. However, due to the simplicity of the circuit, and the ease of updating capacitors can still be used for temporary storage, such as storing the data generated between two clock steps. In such situations, however, the stored data need to be used as quickly as possible to avoid deterioration of charge due to leakage. The synthesis procedure should take care of the situation by properly scheduling the operations so that the data is not required to be stored for more than a few microseconds. Another problem with analog circuits is that they are not easily amenable to be operated multiplexed mode, since analog switches can introduce additional error due to residual charges. Moreover, if the analog data is generated in the form of current, it has to be converted to voltage form for storage. This operation requires extra hardware and may introduce errors due to non-linearity. Therefore analog circuits are to be used sparingly in time-share mode.

The other requirements of the Synthesizer are:

- The preferred mode for implementing any operation is analog. In situations where no analog circuit satisfies the performance requirements, the operation is mapped to digital mode.

- The hardware architecture should be such that analog and digital blocks are well separated. The interfaces between digital and analog blocks should be kept to a minimum.

- Operations that are performed in asynchronous mode should be properly interfaced with the rest of the operations.

- If the operating ranges and signal modes at the input and the output of analog modules are different, suitable circuits should be inserted for interfacing these modules.

3.7.1 Screening Building Block Circuits

Screening is required to select the building block circuits that meet the specifications on tolerable amounts of non-idealities from various operations. It is done by comparing the limits generated by the Analyzer with the behavior models on non-idealities. The behavioral models are generated for circuit parameters and hence they need to be translated for direct comparison with the values generated by the Analyzer. If the circuit performs multiple operations (such as an adder-threshold), the comparison would require forming a composite specification by combining the corresponding terms from the Taylor-series expression and enumerating the result by substituting the computed values for error limits for each operation. For example, if an adder-threshold circuit is used for implementing the addition and thresholding operations in the DFG shown in Fig 3.2, the error specification for the circuit is obtained by adding the corresponding terms in the Taylor-series expression, namely

$$\epsilon_{composite} = \frac{\partial y}{\partial +}\epsilon_+ + \epsilon_f \qquad (3.24)$$

where f is the threshold function and $\epsilon_{composite}$ is the composite expression for the specification of error from the circuit.

As a result of screening, if some analog hardware is found to satisfy the requirements for an operation, the DFG node corresponding to it is marked "analog". Otherwise that node is marked "digital". For data registers, the storage mode is dependent upon the required duration of storage without updating or refreshing the data. Then the modeling procedure given in Section 3.6 is used for calculating the leakage error (ϵ_{cap}) for various capacitor models. Analog storage mode is used if any of them is smaller than the specified limit on error from storing the data. Since ϵ_{cap} is dependent upon the frequency of updating, this would imply selection of digital mode for storage if the frequency is too low or if the data need to be stored permanently without being refreshed.

3.7.2 Reduction of Analog-Digital Interfaces

The presence of analog and digital circuits requires the insertion of additional hardware to convert data from one mode to the other. However the conversion circuits pose many inconveniences to the synthesis procedure. Firstly they are often bulky and hence take away much of the advantage in employing analog circuits for high packing density. Secondly, each conversion is time consuming, and hence can delay the system operation by several clock cycles. This is especially true if the analog-digital interface is in a critical path of operation. Finally, every conversion circuit is made up of some devices that operate in analog mode, and hence these circuits can introduce some non-ideal effects into the system operation. Even if the error due to non-ideal effects is within tolerable limits, their effect can potentially decrease the tolerance of the system to non-idealities from other functional blocks. Because of these adversities it is most advisable to reduce the need for using data converter circuits, which is achieved by minimizing the number of interfaces between analog and digital circuit blocks. In netlist partitioning, this task is akin to minimizing the number of nets that are cut (*i.e,* the nets that are not completely contained in one section). Heuristic techniques are usually employed to execute this task, and most of them perform a series of node movements between the partitioned sections until no more improvement in cuts can be attained [39]. The node movement starts from an initial partitioning of nodes, which in our case is the assignment of the nodes after screening the building blocks. The movement is based upon a measure called *gain* that is calculated for every node which is initially assigned to the analog section. The rest of the nodes are always fixed in the digital section and are never moved. Only the nodes with zero or positive gain are moved at any time, and they are selected in the descending order of their gain values. The gain of a node is contributed by two terms. The first term yields a positive value if there is an immediate reduction in the number of cuts when the node is moved in isolation, or if the node belongs to a group and there is a reduction in cuts by moving all of them together. In an isolated node movement the gain is equal to the change in the number of cuts. For group movement the gain is equal to the reduction in number of cuts divided by the number of nodes in the group. The group is formed by all the nodes that are connected to a common net but the movement of each of them in isolation causes neither an increase nor a decrease in the number of cuts. The other gain term is calculated for possible change in circuit sizes. The gain is +0.5 if the movement is from digital to analog, and it is -0.5 vice versa. Usually node movement heuristics are guided by a specification on the maximum and minimum number of nodes to be accommodated in each section, in order to prevent all the nodes from one section getting moved eventually to the other section. In analog-digital partitioning it is difficult to assign these limits, therefore the gain due to change of circuit sizes helps in applying some "brake" to unrestricted node movements that result in erosion of nodes from the analog section.

3.7.3 Scheduling and Allocation

A high-level description of interconnected hardware blocks is synthesized by *scheduling* the operations to different clock steps and *allocating* specific instances of circuit blocks for executing the operations. Optimum scheduling and allocation is an NP-hard problem [40], therefore many heuristics have been proposed to perform these tasks in polynomial time [2]. There is no preferred order in which these two tasks are to be executed, and some heuristics perform them simultaneously [41, 42]. Scheduling requires that the order of execution of operations is known. The order may be determined by As-Soon-As-Possible (ASAP) or As-Late-As-Possible (ALAP) schemes. In the ASAP scheme the operations are assigned to the earliest available clock step, without violating the precedence of operations. Since one of the requirements for scheduling analog operations is that the duration of temporary data storage be kept minimum, analog operations are ordered using ASAP ordering. In the ALAP ordering, the operations are delayed until their outputs are required by succeeding operations. ALAP ordering is preferred for achieving more flexibility in scheduling [24], and hence digital operations are ordered using this scheme. The objective of scheduling is to perform the operations using as few clock steps as possible. Scheduling of analog operations is not as straightforward as that of digital operations. For example, since signals in current form cannot be stored, all analog operations that use current data and are performed in sequence need to be executed at the same clock step. If the analog-digital interfaces are loosely coupled, then it is possible to schedule analog and digital operations separately, with each interface modeled as a sequential synchronous data I/O with fixed communication rate [43]. A typical ANN system with the operations mapped to analog and digital modes is characterized by the presence of many interfaces between the two sections. Therefore the partition is tightly coupled and hence the techniques suggested in [43] can result in long waiting times for the operations in both the sections. Therefore we use an approach whereby the operations are integrated into a general scheduling and allocation framework. This approach uses ASAP ordering for analog operations, assigns analog operations to be used in multiplexed mode only if the operands do not change or the tolerance to errors is high, and schedules a sequence of analog operations using current signals to one clock step but lets them be assigned to separate clock steps if the interfacing data is in voltage form. The low hardware cost of analog operations is fully utilized by adopting simultaneous scheduling and allocation to incrementally improve the value of a cost function given by the equation:

$$Total\ Cost\ C\ =\ \sum_i c_{hw}(i) \cdot N_{hw}(i)\ +\ c_{clk} \cdot T_e \qquad (3.25)$$

where $c_{hw}(i)$ is the cost of hardware unit i, $N_{hw}(i)$ is the number of instances of unit i that are used, c_{clk} is the cost per clock step, and T_e is the execution time in number of clock steps. The overall resource management is effected by following a technique called *selective sequentialization* of operations that are otherwise ordered to be executed concurrently. We have already seen at the beginning of the section that sequentialization of operations and functional pipelining can result in efficient trade-of between hardware cost and length of the execution cycles. By this technique, the group of concurrent operations that offer the best sequentialization result is selected based upon a quantity called *reduction coefficient*. The reduction coefficient is calculated by considering all the operations in a group together. It is an indicator of savings in hardware at the expense of increased execution time, due to sequentialization of operations in that group. The reduction coefficient (rc) for a group j is calculated as:

$$rc_j = (c_{hw}(j) \cdot n_{hw}(j) + rc\_fan_j - c_{clk} \cdot n_{hw}(j)) \times sf \quad (3.26)$$

where sf is the sequentialization factor. It is set to a value lying between 0 and 1, with 1 indicating full sequentialization. $n_{hw}(j)$ is the number of concurrent operations in group j and $c_{hw}(j)$ is the hardware cost for executing each of them. The first term signifies the reduction in hardware cost due to sequentialization of operations in the group. The second term indicates additional reduction in hardware (without increasing the execution time) due to possible sequentialization of operations which are connected to the edges fanning out from operators of the selected group. This additional sequentialization is possible since these operations are dependent upon the data from the sequentialized group j and hence there is no advantage in operating them in parallel. It is calculated using the equation:

$$rc\_fan_j = \sum_{k=1}^{f} \frac{n_{hw}(k)}{fan_j(k)} \times c_{hw}(k) \quad (3.27)$$

where f is the number of different operational classes (operations of the same type and having identical specifications belong to one class) connected to the fanned-out edges from the operations in group j, and $fan_j(k)$ is the number of operators in each of those classes. The quantity $fan_j(k)$ is required to ensure pipelining of operations in group k with those in group j, without causing any additional increase in the number of clock steps. The third term in (3.26) represents the increase in execution time. Since the increase in time is detrimental, its sign is negative. Equation (3.26) shows that by setting different values to c_{hw} and c_{clk}, the emphasis can be shifted between reduction in hardware and preservation of parallelism.

The scheduling and allocation step starts from an initial ordering of operations and by forming groups of concurrent operations. These operations are scheduled for maximum concurrency and the value of the cost function C given in Equation (3.25) is calculated for the resulting hardware allocation. This cost is iteratively improved by sequentializing the group with the highest reduction coefficient. The iteration stops when all the groups have been sequentialized or when there is no more reduction in cost due to sequentialization. After every iteration the operations are re-scheduled and hardware resources are re-allocated for pipelining.

3.7.4 Generation of Final Interconnect

The scheduling and allocation procedures described in the previous section are for synthesizing a hardware interconnection scheme in which the operations are synchronized to different time steps of a system clock. But the learning in many ANN systems is performed in asynchronous mode, and in those situations it is hard to sequentially order the operations and tie each of them to a clock step. A different strategy is required to handle sequences of operations that are executed asynchronously and to formulate a schedule so that these operations are properly interfaced with the rest of the system operating in synchronous mode.

Asynchronously operating node paths are introduced in the DFG description by forming register-less loops with the required nodes. In the absence of registers to store data in a cyclic path or loop, the order of execution of operations in that path is indeterminate, and hence they cannot be assigned to successive steps of the system clock. In our synthesis environment, these operations are bunched together and are assigned to a "macro" clock step. The duration of the macro step is specified as an integer multiple of the duration of a basic clock step. The multiplication factor is set by the designer by estimating the settling time for the signals in that path. Since multiplexing is not possible in such a scheme, separate hardware units are allocated to each of the operators. Operations in an asynchronous path are not separated by clock steps. Each operator is allocated a separate hardware unit. The assignment of macro clock steps to asynchronous paths helps in interfacing them with the rest of the nodes in the DFG, without causing any conflicts in data requirements.

The other steps to be executed before the final hardware description is generated include optimization of hardware by employing multi-function blocks, and insertion of buffering circuits to interface analog blocks with incompatible I/O modes and ranges.

3.8 DESIGN EXAMPLES

The use of the design environment for synthesizing mixed analog/digital hardware is illustrated with the help of four examples from three different ANN systems.

Mixed Analog/Digital Design Automation 123

The first two examples are for applications that use the Multi-Layer Perceptron (MLP). These ANNs employ the Back-Propagation algorithm for learning. We start with the smaller of the two, which is the 2-input Exclusive-OR problem [8]. A three-layer network is used for this problem, with 2 neurons at the input layer, 2 neurons at the hidden layer, and one neuron at the output layer. The statistical data on various random variables are given in Table 3.2. The distributions of signals emerging from the threshold (the *tanh* function is used here) and differential of threshold (nodes 23, 25, and 42) operations, is obtained by approximating them as functions of synaptically weighted inputs, which in turn are normally distributed. The network is trained with all the four possible

Table 3.2 Statistical data on random variables

Random Variable	Distribution	Range	Mean	Variance
Input Data	Smooth bi-modal	-1 to +1	0	0.05
Weights and Biases	Uniform	-1 to +1	0	0.03333
Threshold (output of)	Steep bi-modal	-1 to +1	0	1.0
Difftl. of Threshold (output of)	-	0 to 1	0.056	0.037

sign combinations of the input signals, and by forcing the activations from the output layer to match the expected value (+1 if both inputs have same sign and -1 if they are different). The DFG for this example is shown in Figure 3.6. The input signals to the ANN system enter through the nodes numbered 1 and 2. The synaptic weights and neuron biases are stored and updated at the nodes 3, 4, 5, 6, 7, 8, 27, 28, and 29. The DFG is evaluated by the Analyzer using the equality constraints obtained from Equations (3.7), (3.9), (3.11), and (3.13) and by applying the inequality constraint in (3.14) at the output of the nodes 19, 20, 21, 22, 35, and 36, with the value of c set to 0.9. The limits on error from various operations, generated by the Analyzer, are listed in Table 3.3. Table 3.4 summarizes the behavioral models of analog building block circuits. The variance values in the fourth column are for the parameters (given in the second column), and the values in the last column are calculated by applying the translation procedure. Table 3.5 shows the DFG nodes that are initially mapped to analog operation. This table is obtained by comparing the second column of Table 3.3 with the values in the last column of Table 3.4. All the nodes that store weights and neuron biases are mapped to the digital block to satisfy the requirements for permanent storage. In order to minimize the number of analog-digital interfaces, the nodes numbered 23, 25, 35, 36, 37, and 38 are moved from the analog section to the digital section. The hardware cost

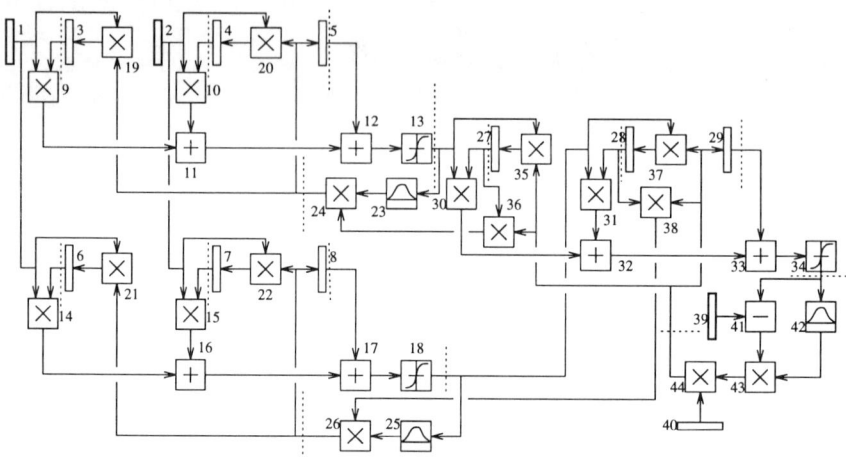

Figure 3.6 DFG for the 2-input Exclusive-OR problem

Table 3.3 Summary of upper limits generated by the analyzer on variance of error due to hardware related non-idealities

Operation	Limits from Analyzer	Nodes in DFG
Multiplication (forward)	2.25×10^{-3}	9,10,14,15,30,31
Addition (forward)	2.25×10^{-3}	11,12,16,17,32,33
Tanh function	3.6×10^{-4}	13,18,34
Difference	1.8×10^{-4}	41
Multiplic (back-prop error)	3.6×10^{-6}	24,26
Multiplication (learning error)	2.81×10^{-4}	36,38,43,44
Multiplication (wt updt-hidden layer)	3.6×10^{-6}	19,20,21,22
Multiplication (wt updt-output layer)	1.23×10^{-3}	35,37
Differential of Tanh (hidden layer)	4×10^{-5}	23,25
Differential of Tanh (output layer)	4×10^{-4}	42
Weight storage	1×10^{-3}	3,4,5,6,7,8,27,28,29

of various building blocks for synthesis are shown in Table 3.6. The initial ordering is made by distributing the operations into 17 groups, and they were

Table 3.4 Summary of behavioral models of analog building block circuits

Circuit	Non-ideal Parameter	Mean	Variance	Translated Variance
Gilbert Multiplier	transcond. gain	≈ 0	7.96×10^{-14}	1.0×10^{-3}
Quarter-Square Multiplier	voltage gain	≈ 0	2.77×10^{-4}	2.77×10^{-4}
Adder-threshold (op-amp)	*compound*	≈ 0	-	4.0×10^{-4}
Differential amp (threshold)	voltage gain	≈ 0	2×10^{-4}	2×10^{-4}
MDAC (5 bits)	transcond. gain	≈ 0	2.57×10^{-14}	2.98×10^{-3}
Difference circuit (diff. pairs)	voltage gain	≈ 0	1.5×10^{-4}	1.5×10^{-4}
MOS Resistor	conductance	≈ 0	6.2×10^{-13}	6.2×10^{-3}

Table 3.5 Mapping DFG nodes to analog building block circuits

Nodes	Building Block Circuits
9,10,14,15,30,31	Gilbert Multiplier, Quarter-Sq. Multiplier, Multiplying DAC (compound)
35,36,37,38,43	Quarter-Sq. Multiplier
11,12,16,17,32,33	Op-amp current adder, Adder-Threshold (compound)
13,18,34	Thresholding circuit, Adder threshold (compound)
41	Difference circuit, Quarter Sq. Multiplier *(compound operation, combined with mult. operation at 43)*
23,25,42	Quarter-Square Multiplier[b]
1,2,39	Capacitor storage
40,41	MOS Resistor *(multiplication with a constant)*

[b] If $y = \tanh(x)$, the differential can be written as $(1-y) \cdot (1+y)$.

scheduled to operate in 11 clock steps for maximum concurrency. The initial synthesis cost for this schedule was 2383, which was reduced to 2231 by sequentializing the group containing the multiplication operations at nodes 19, 20, 21, and 22 with the value of sf in Equation (3.26) set to 0.5.

Table 3.6 Hardware and Clock Step costs

Building Block	Cost
Analog multiplier (Gilbert, Quarter-Sq.)	10
Constant multiplication (MOS resistor)	1
Analog Adder	1
Analog Threshold (*tanh*)	15
Analog differential of threshold	10
Digital multiplier	$1 \times bits^2$
Digital Adder	$1.25 \times bits$
Digital differential of threshold	$1.25 \times bits^2$
D/A converter	0.25×2^{bits}
A/D converter	$0.25 \times 2^{bits} + 16$
Clock	150

The second MLP example is for recognizing printed digits from 0 to 7. The digits are printed on a 5x4 pixel plane. There is a one-to-one mapping between a pixel and a neuron of input layer, hence there are 20 neurons at that layer. The hidden layer has 6 neurons and the output layer has 3 neurons. Learning is performed to achieve a unique one-to-one mapping between the digit appearing at the input plane and a 3-bit code produced from the activations of the output layer neurons. The Back-Propagation algorithm is used for learning the task. At the input plane a dark pixel is indicated by a value which is close to +1 and a light pixel by one close to -1. The distribution of random input variables is as shown in Table 3.2. The DFG for this problem is given in Figure 3.7. The error limits generated by the Analyzer are shown in Table 3.7. Using the error limits, digital circuits are assigned for weight updating operations and for the operations that calculate back-propagated error. The rest of the operations are implemented using analog circuits. The initial ordering of operations resulted in a scheduling scheme that used 13 clock steps to complete one system cycle. Two configurations are generated from this ordering, by applying selective sequentialization. The first one was generated with emphasis placed on reducing hardware cost and the second one was for using the minimum number of clock steps. For the first scheme the final schedule and hardware allocation

Mixed Analog/Digital Design Automation

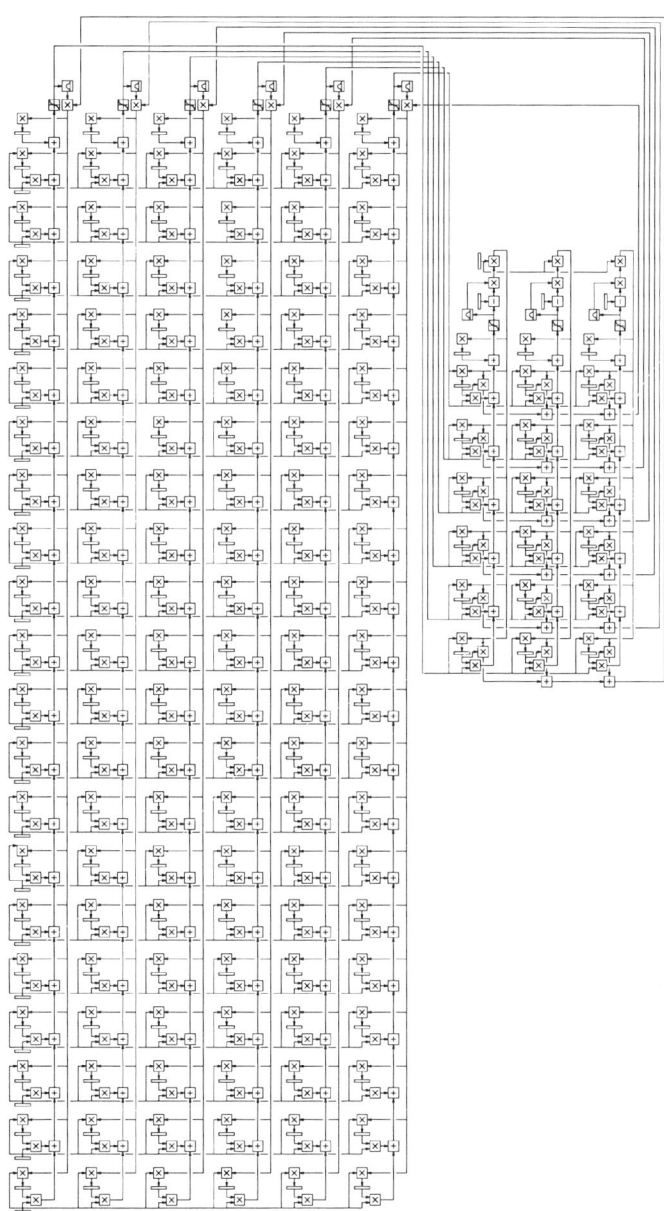

Figure 3.7 DFG for the digits recognition problem with learning

Table 3.7 Upper limits generated by the analyzer on variance of error, and comparison of the values with simulated results

Operation	Limits from Analyzer
Multiplication(forward)	2.25×10^{-3}
Addition (forward)	2.25×10^{-3}
Tanh function	1.17×10^{-3}
Difference	8×10^{-5}
Multiplication(back-prop error)	1×10^{-8}
Multiplication(learning error)	3×10^{-4}
Multiplication (weight updt-hidden layer)	1×10^{-8}
Addition (back-prop error)	2.25×10^{-3}
Multiplication (weight updt-output layer)	3×10^{-4}
Differential of Tanh (hidden layer)	1×10^{-6}
Differential of Tanh (output layer)	3×10^{-4}
Weight storage	1×10^{-3}

are shown in Figure 3.8. The total cost was minimized from 18,252 to 6863 for this scheme. In order to generate the second scheme the cost of a unit clock step was increased from 150 to 500. The final schedule and hardware allocation, shown in Figure 3.9, was generated by minimizing the total cost from 24,102 to 11,991. The third application example uses a Hopfield Network to implement a 4-bit A/D converter [36]. A 4 neuron network is used for this example. The input to the network is generated from the analog input to the converter using the relation:

$$I = ax + b \qquad (3.28)$$

where a and b are constant vectors and x is the analog input. The T matrix, and the vectors a and b for this example are given below:

$$T = \begin{bmatrix} 0 & -2 & -4 & -8 \\ -2 & 0 & -8 & -16 \\ -4 & -8 & 0 & -32 \\ -8 & -16 & -32 & 0 \end{bmatrix}, \quad a = \begin{bmatrix} 1 \\ 2 \\ 4 \\ 8 \end{bmatrix}, \quad b - \begin{bmatrix} 0.5 \\ 2 \\ 8 \\ 32 \end{bmatrix} \qquad (3.29)$$

Figure 3.10 shows the DFG for this application of the Hopfield network. The error limits generated by the Analyzer, after applying the constraint shown in Equation (3.18) to the outputs from nodes 40, 52 (the two least significant bits) of the DFG, are shown in Table 3.8. The corresponding values for other bits

Mixed Analog/Digital Design Automation

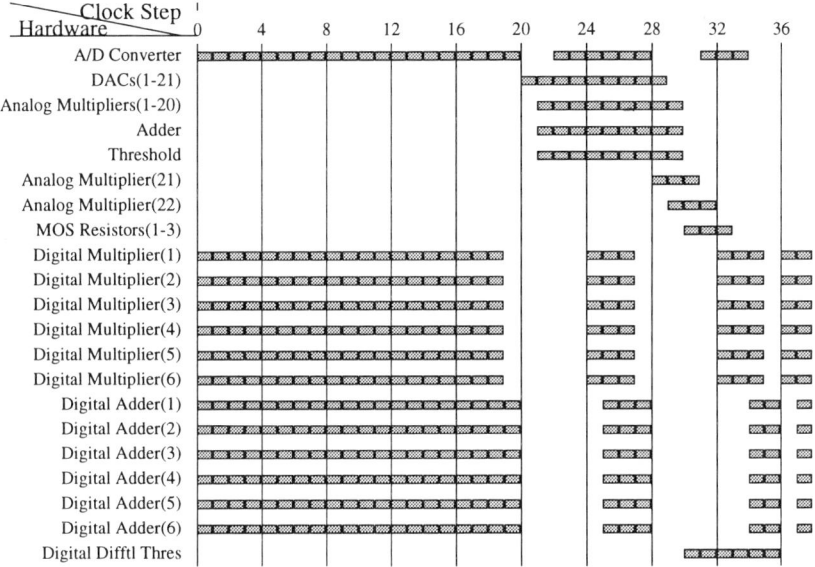

Figure 3.8 Schedule and allocation chart for the digit recognition problem

Table 3.8 Error limits for hardware implementation of 4-bit A/D converter

Operation	Error bound (LSB)	Error bound (next LSB)	Nodes in DFG (LSB)
Multiplication	0.09	0.4	30,31,32,33,34,40
Addition	0.09	0.4	35,36,37,38,39
T	0.06	0.319	2,3,4,5
x	0.09	0.4	1
a	0.011	0.05	6
b	0.09	0.4	7
R	0.085	0.1	8

were found to be significantly higher. The limits indicate that all the nodes of the DFG can be implemented using analog circuits. Digital registers were used for storing the T matrix, and the vectors a and b. The interconnection of the nodes in the DFG indicate that the nodes numbered from 30 through 77

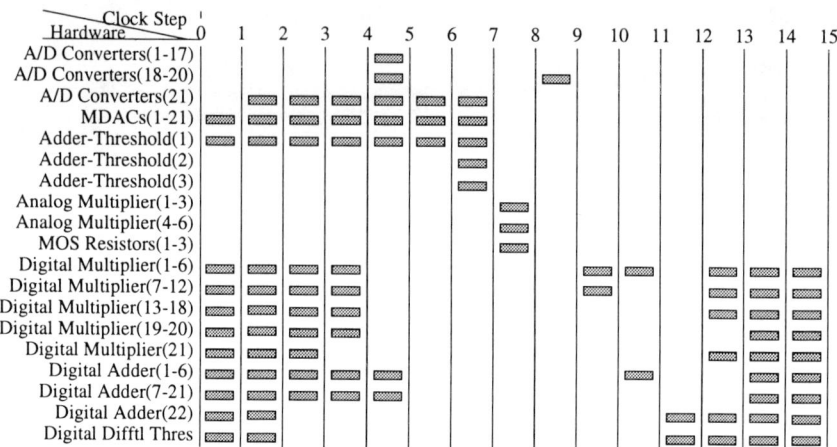

Figure 3.9 Schedule and allocation chart for the synthesized configuration with emphasis placed on reduction of number of clock steps

are to be operated in asynchronous mode. The allocation of this scheme used 24 analog multipliers operating in parallel. During the merge-step, 20 of these multiplications were merged with the D/A conversions that follow T and b registers, resulting in the use of as many MDACs. The merge step also combined the add operations at nodes 35,36,37,38, and 39 to one circuit. Similar optimization is also applied at the other nodes performing add operation. Detecting that current signals are multiplied with a constant at nodes 40, 52, 64, and 76, the optimization also step merged the multiplication operations with the storage operations at the nodes 8, 15, 22, and 29 and allocated simple resistors that have one end grounded.

For the last example a Deterministic Boltzmann ANN with 8 input neurons and 2 neurons each at the hidden and the output layers is used to classify 36 patterns into 4 different classes. The DFG for this network is shown in Figure 3.11. Even though it is a multi-layer ANN, for this application each neuron is connected to all other neurons. The error limits generated by the Analyzer by applying the constraint in Equation (3.21) and with the values of k_1 and k_2 set to 0.001 (for accurate matching), are tabulated in Table 3.9. Using these results and the behavioral models, Gilbert multipliers are used for forward path multiplications and Quarter-Square multipliers are used for weight computation multiplications. Intermediate values are stored in capacitors and differential pairs are used for the difference operations. Current summing amplifiers are used for addition. A Gilbert multiplier with one input held at a constant value is

Mixed Analog/Digital Design Automation 131

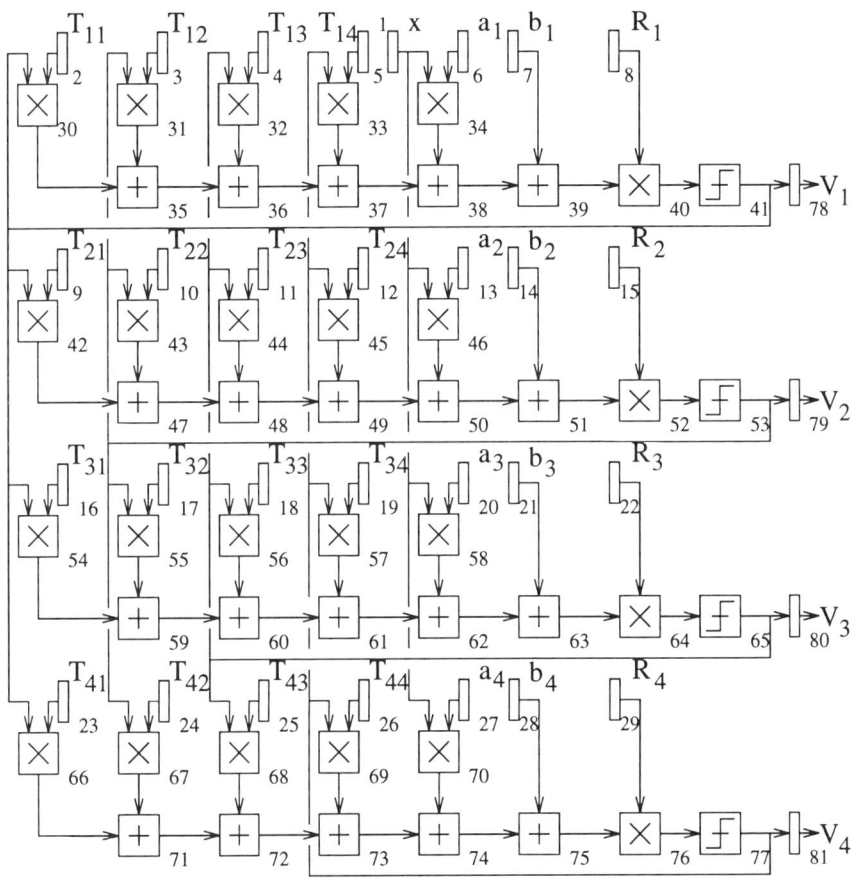

Figure 3.10 Hopfield Network for the 4-bit A/D converter example

used for implementing the thresholding operation. For this application, training data are interspersed with the recognition patterns, and hence the weights are constantly updated. Therefore capacitors are used for weight storage.

The hardware was synthesized to operate in three phases - one each for the unclamped and clamped phases and one for weight update. The operations in the unclamped and clamped phases are performed asynchronously, and hence a separate circuit is assigned to each DFG node. But unclamped and clamped phases are non-overlapping, and hence the same circuit blocks are used for

Table 3.9 Error bounds for hardware implementation of pattern classifier

Operation	Error bound
Multiplication (forward)	1.45×10^{-3}
Addition (forward)	1.45×10^{-3}
Multiplication (wt. updt-hidden&output)	4×10^{-4}
Multiplication(wt. updt-input layer)	1.26×10^{-3}
Difference (hidden and output layers)	4×10^{-4}
Difference (input layer)	1.26×10^{-3}
Sigmoid	4×10^{-4}
Weight storage	1×10^{-3}

performing both the phases. The synthesized hardware contains 132 Gilbert multipliers, 132 Quarter-Square multipliers, 132 differential pairs, 12 current summing op-amps, and 132 capacitors and weight updating circuits.

3.9 CONCLUSIONS

In this chapter we have introduced an environment for automating the design of mixed analog/digital VLSI hardware for ANN applications. The environment helps to systematically investigate many implementation choices and hence to generate prototype designs with shorter turn-around times. Unlike other synthesis programs this environment uses a unified approach for automating the designs, but at the same time specifically addresses the conflicting requirements of analog and digital operations. The non-ideal effects from analog circuit operations are studied quantitatively using the Analyzer tool and specifications on tolerable limits on error due to these effects are generated. The Synthesizer tool uses these specifications to select suitable circuit blocks. The synthesized hardware is predominantly analog. But when digital blocks need to be used, for example to implement learning operations that require high computational accuracy, the Synthesizer finds an optimum mixture of analog and digital operations. It generates a high-level interconnection description of building block hardware by simultaneously scheduling both analog and digital operations. The selective sequentialization method for scheduling and allocation not only addresses the concurrency of ANN operations, but also helps to synthesize the hardware with varying emphasis on reduction of hardware cost

Mixed Analog/Digital Design Automation

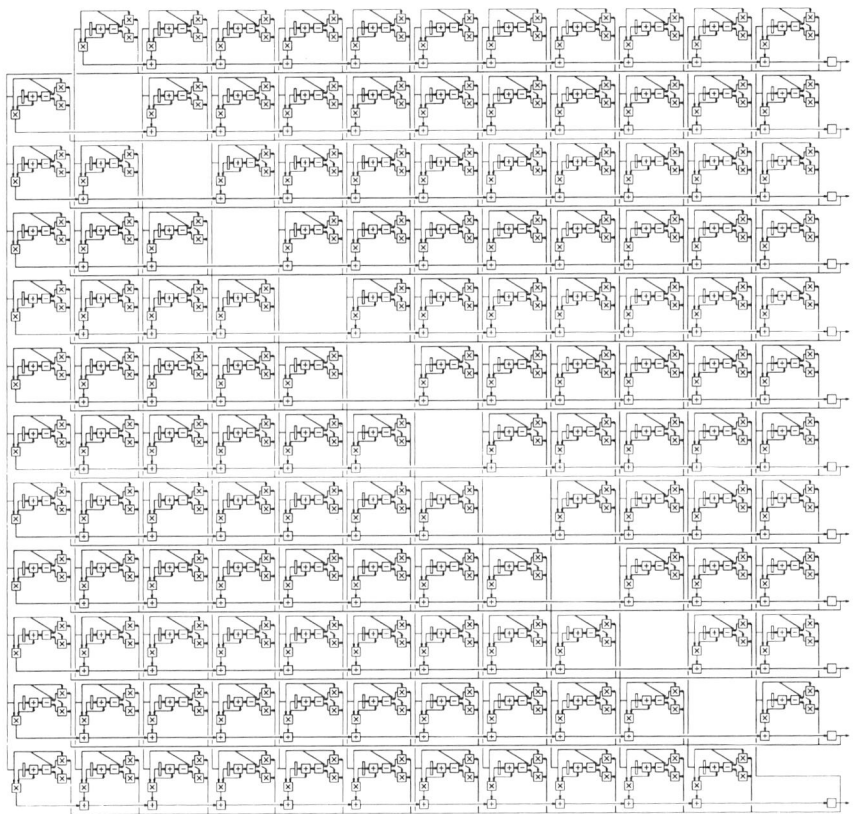

Figure 3.11 DFG for Pattern Classification using a Deterministic Boltzmann ANN

or improving the performance throughput. The application example that uses a Multi-Layer Perceptron to solve the digit recognition problem illustrates this feature. The automation methodology is very general and so far we have used it for designing Multi-Layer Perceptrons, Hopfield Networks, and Deterministic Boltzmann machines. It is hoped that the environment caters to hardware designs for application problems using other ANN systems as well. It would also be worthwhile to examine how the environment can be extended to automating the design of mixed analog/digital hardware for other applications, such as telecommunication, instrumentation, and signal processing.

REFERENCES

[1] E. Sanchez-Sinencio and C. Lau, eds., *Artificial Neural Networks: Paradigms, applications, and hardware implementations.* New York, NY: IEEE Press, 1991.

[2] R. A. Walker and R. Camposano, eds., *A Survey of High-Level Synthesis Systems.* Boston, MA: Kluwer Academic Publishers, 1991.

[3] B. Haroun and E. Torbey, "Synthesis of Multiple Bus/Functional Unit Architectures Implementing Neural Networks," in *Proc. Int'l Conf. on Computer Design*, (Cambridge, MA), pp. 174–178, 1992.

[4] H. P. Graf, L. D. Jackel, and W. E. Hubbard, "VLSI Implementation of a Neural Network Model," *IEEE Computer Magazine*, pp. 41–49, March 1988.

[5] J. N. Babanezhad and G. C. Temes, "A 20-V Four-Quadrant CMOS Analog Multiplier," *IEEE Journal of Solid-State Circuits*, vol. 20, no. 6, pp. 1158–1168, December 1985.

[6] S. P. Eberhardt, T. Duong, and A. Thakoor, "Design of Parallel Hardware Neural Network Systems from Custom Analog VLSI 'Building Block' Chips," in *Proc. Int'l Joint Conf. on Neural Networks*, vol. II, (San Diego, CA), pp. 183–190, June 1989.

[7] W.-C. Fang, B. J. Sheu, O. T.-C. Chen, and J. Choi, "A VLSI Neural Processer for Image Data Compression Using Self-Organization Networks," *IEEE Trans. on Neural Networks*, vol. 3, pp. 506–518, May 1992.

[8] J. S. Pena-Finol and J. A. Connelly, "A MOS Four-Quadrant Analog Multiplier Using the Quarter-Square Technique," *IEEE Journal of Solid-State Circuits*, vol. 22, no. 6, pp. 1064–1073, December 1987.

[9] F. M. A. Salam, N. Khachab, M. Ismail, and Y. Wang, "An Analog MOS Implementation of the Synaptic Weights for Feedback Neural Nets," in *Proc. IEEE Int'l Symposium on Circuits and Systems*, (San Diego, CA), pp. 1223–1226, May 1989.

[10] D. E. Rumelhart, G. Hinton, and R. J. Williams, "Learning internal representations by error propagation," in *Parallel Distributed Processing : Explorations in the Microstructure of Cognition (Vol. 1)* (D. E. Rumelhart and J. L. McClelland, eds.), Cambridge, MA: M. I. T. Press, 1986.

[11] J. A. Lansner and T. Lehmann, "An Analog CMOS Chip Set for Neural Networks with Arbitrary Topologies," *IEEE Trans. on Neural Networks*, vol. 4, no. 3, pp. 441–444, May 1993.

[12] Y. Wang, "Analog CMOS Implementations of Backward Error Propagation," in *Proc. IEEE Int'l Conference on Neural Networks*, (San Francisco, CA), pp. 701–706, March 1993.

[13] C. A. Mead, *Analog VLSI and Neural Systems*. Reading, MA: Addison-Wesley, 1989.

[14] S. Y. Foo, L. R. Anderson, and Y. Takefuji, "Analog Components for the VLSI of Neural Networks," *IEEE Circuits and Devices Magazine*, pp. 18–26, May 1990.

[15] R. L. Shimabukuro, P. A. Shoemaker, and M. E. Stewart, "Circuitry for Artificial Neural Networks with Non-Volatile Analog Memories," in *Proc. IEEE Int'l Symposium on Circuits and Systems*, (Portland, OR), pp. 1217–1220, May 1989.

[16] M. Holler, S. Tam, H. Castro, and R. Benson, "An Electrically Trainable Artificial Neural Network (ETANN) with 10240 "Floating Gate" Synapses," in *Proc. Int'l Joint Conf. on Neural Networks*, vol. II, (San Diego, CA), pp. 191–196, June 1989.

[17] J. J. Hopfield, "The Effectiveness of Analogue 'Neural Network' Hardware," *Network*, vol. I, pp. 27–40, 1990.

[18] H. A. Castro, S. M. Tam, and M. Holler, "Implementation and Performance of an Analog Non-Volatile Neural Network," in *80170NX Neural Network Technology and Applications* (M. Holler, ed.), Santa Clara, CA: Intel Corporation, 1992.

[19] B. E. Boser and E. Sacklinger, "An Analog Neural Network Processor with Programmable Network Topology," in *Digest of Technical Papers, IEEE Int'l Solid-State Circuits Conf.*, (San Francisco, CA), pp. 184–185, February 1991.

[20] C. Schneider and H. C. Card, "Analog CMOS Deterministic Boltzmann Circuits," *IEEE Journal of Solid-State Circuits*, vol. 28, no. 8, pp. 907–914, August 1993.

[21] C. Schneider and H. C. Card, "Analog CMOS Contrastive Hebbian Networks," in *SPIE Applications of Artificial Neural Networks III*, vol. 1709, (Orlando, FL), pp. 726–735, 1992.

[22] E. Berkcan and F. Yassa, "Towards Mixed Analog/Digital Design Automation: A Review," in *Proc. IEEE Int'l Symposium on Circuits and Systems*, (New Orleans, LA), pp. 809–815, May 1990.

[23] H. Chang et al., "A Top-Down Constraint-Driven Design Methodology for Analog Integrated Circuit," in *Proc. IEEE Custom Integrated Circuits Conf.*, (Boston, MA), pp. 8.4.1–8.4.6, 1992.

[24] B. S. Haroun, *VLSI Architectural Synthesis for DSP Custom Applications*. PhD thesis, Department of Electrical Engineering, University of Waterloo, Waterloo, Ont., 1989.

[25] R. C. Frye, E. A. Reitman, and C. W. Wang, "Back-Propagation Learning and Nonidealities in Analog Neural Network Hardware," *IEEE Trans. on Neural Networks*, vol. 2, no. 1, pp. 110–117, January 1991.

[26] J. B. Lont and W. Guggenbuhl, "Analog CMOS Implementation of a Multilayer Perceptron with Nonlinear Synapses," *IEEE Trans. on Neural Networks*, vol. 3, no. 3, pp. 457–465, May 1992.

[27] B. Dolenko and H. C. Card, "The Effects of Analog Hardware Properties on Backpropagation Networks with On-Chip Learning," in *Proc. Int'l Conf. on Neural Networks*, (San Francisco, CA), pp. 110–115, March 1993.

[28] S. Kullback, *Information Theory and Statistics*. New York: Wiley, 1959.

[29] M. R. Walker and L. A. Akers, "Information-Theoretic Analysis of Finite Register Effects in Neural Networks," in *Proc. Int'l Joint Conf. on Neural Networks*, vol. II, (Baltimore, MD), pp. 666–669, June 1992.

[30] R. Spence and R. S. Soin, *Tolerance Design of Electronic Circuits*. Reading, MA: Addison-Wesley, 1988.

[31] Y. Kumar and J. P. Knight, "Automatic Word Length Determination in Behavioral Synthesis of Digital Circuits," in *Proc. Canadian Conf. on VLSI*, (Ottawa, Ont.), pp. 6.2.1–6.2.8, October 1990.

[32] J. L. Holt and J.-N. Hwang, "Finite Precision Error Analysis of Neural Network Hardware Implementations," *IEEE Trans. on Computers*, vol. 42, no. 3, pp. 281–290, March 1993.

[33] A. Achyuthan, *VLSI Hardware Synthesis of Artificial Neural Networks*. PhD thesis, Department of Electrical and Computer Engineering, University of Waterloo, Waterloo, Ont., Canada, 1993.

[34] A. Papoulis, *Probability, Random Variables and Stochastic Processes.* New Jersey: McGraw-Hill Inc., 1991.

[35] J. J. Hopfield and D. W. Tank, "'Neural' Computation of Decision Optimization Problems," *Biological Cybernetics*, vol. 52, no. 3, pp. 141–152, July 1985.

[36] M. Vidyasagar, "An Analysis of the Flows of Neural Networks with Linear Interconnections," in *Proc. IEEE Int'l Joint Conf. on Neural Networks*, vol. III, (San Diego, CA), pp. 523–528, June 1990.

[37] N. A. Group, *NAG Fortran Library Mark 15.* Oxford, UK: NAG, 1992.

[38] A. Achyuthan and M. I. Elmasry, "A Methodology for Generating Behavioral Specifications of Analog Hardware for Artificial Neural Network Implementations," in *SPIE Applications of Neural Networks IV*, (Orlando, FL), April 1993.

[39] C. M. Fiduccia and R. M. Mattheyses, "A Linear-Time Heuristic for Improving Network Partitions," in *Proc. 19th Design Automation Conf.*, (Las Vegas, NV), pp. 175–181, June 1982.

[40] M. R. Garey and D. S. Johnson, *Computers and intractability: a guide to the theory of NP-completeness.* San Francisco, CA: W. H. Freeman, 1979.

[41] P. G. Paulin and J. P. Knight, "Force-Directed Scheduling for the Behavioral Synthesis of ASIC's," *IEEE Trans. on Computer-Aided Design*, vol. 8, no. 6, pp. 661–679, June 1989.

[42] S. Devadas and A. R. Newton, "Algorithms for hardware allocation in data path synthesis," *IEEE Trans. on Computer-Aided Design*, vol. 8, no. 6, pp. 661–679, June 1989.

[43] C. H. Gebotys, *A Global Optimization Approach to Architecural Synthesis of VLSI Digital Synchronous System with Analog and Asynchronous Interfaces.* PhD thesis, Dept. of Electrical and Computer Eng., University of Waterloo, Waterloo, Ont., 1991.

4

A COMPACT VLSI IMPLEMENTATION OF NEURAL NETWORKS

Liang-Yong Song, Anthony Vannelli and Mohamed I. Elmasry

4.1 INTRODUCTION

The effectiveness of an Artificial Neural Network (ANN) algorithm strongly depends on the hardware that executes it. Because of the similarity of neural and electronic circuitry, it is possible to emulate many neural functions in VLSI [1]. Analog remains potentially advantageous over digital for low-precision processing [2]. Fortunately, ANNs are more tolerant of low-precision components than conventional computation systems. In particular, ANNs with learning can learn weights which compensate for component variations [3]. Analog implementations of common components, such as multipliers and adders, are much more silicon area efficient than their digital equivalents in the low-precision applications. Therefore analog implementation of ANNs could be more efficient than digital implementation.

Although there are a lot of ANN algorithms available, for variety of applications, the most popular implementation of the ANN algorithms is an emulation by a digital computer. It cannot meet the requirements in real-life applications. Only with the ANN hardware implementations can ANN applications be possible. Since most of the ANN algorithms study is isolated from ANN implementation research, most of the learning algorithms are too complicated to implement as special hardware. Some modified rules for hardware implementation have been proposed [4,5].

On-chip learning neural networks not only realize a fast learning process and a low cost of information processing system, but they also realize compensation for any inferior factors that depend on VLSI process or circumstances. In this chapter, we describe a compact implementation of neural networks.

We assume that the inputs and outputs of neurons are digital. This assumption is reasonable in some applications, such as pattern recognition application. Under this assumption, the weighted interconnection can be approximated by switching analog current which is much simpler and more area-efficient circuit implementation. A modified learning rule has been adapted to realize a circuit implementation. The key feature of the modified learning rule is that its implementation needs less circuitries. Therefore a large scale network can be implemented on a chip. Section 4.2 describes the circuit design of the building blocks to implement the basic neural computations. The circuit implementation example is given in Section 4.3. The expanding of the network with multiple chips is discussed in Section 4.4. Conclusion is presented in Section 4.5.

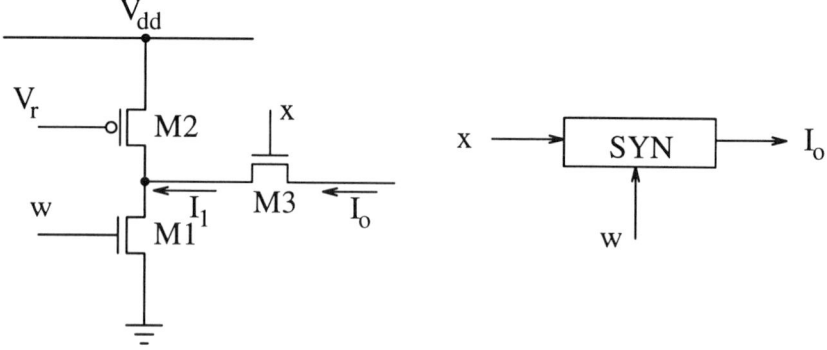

Figure 4.1 The synapse circuit

4.2 THE BUILDING BLOCKS

4.2.1 The Synapse Circuit

If the input and output of a neuron are digital values, the weighted interconnection $w_{ji}x_i$ can be efficiently approximated by switching the analog current. This weighted interconnection can be implemented by the synapse circuit shown in Figure 4.1. The output current I_o is obtained by switching analog current I_1. Let

$$I_{M1} \equiv f(w), \qquad (4.1)$$

then

$$I_1 = I_{M1} - I_{M2}$$

$$= f(w) - I_{M2}$$
$$\equiv f_1(w). \quad (4.2)$$

Transistor $M2$ generates the bias current which is controlled by reference voltage V_r, so the current I_1 can be changed from negative to positive when the weight voltage w changes from 0 to 5v. Transistor $M3$ is used as a switch which is controlled by x. So the output current I_o can be written as

$$I_o = f_1(w) f_2(x), \quad (4.3)$$

where f_1, f_2 are nonlinear function in general. Figure 4.2 shows the synapse circuit characteristic simulated by HSPICE. Simulations throughout this chapter are based on the Northern Telecom 1.2μm CMOS technology [6]. If x is digital signal, when x is high, $f_2(x)$ is one, when x is low, $f_2(x)$ is zero. In this case, Equation (4.3) can be expressed by

$$I_o = k f_1(w) x. \quad (4.4)$$

If the weight is defined by
$$wl = f_1(w), \quad (4.5)$$
or
$$w = f_1^{-1}(wl), \quad (4.6)$$

where f_1^{-1} is the inverse function of function f_1, then Equation (4.4) can be rewritten as

$$I_o = kwlx. \quad (4.7)$$

So Equation (4.7) is the weighted interconnection operation. Note that the weight wl has to be dynamically updated according to the learning rules during learning procedure. The inverse function f_1^{-1} has to be implemented to update the gate voltage w in circuit implementation. This will increase the complexity of circuit implementation. In other word, we may use Equation (4.4) to approximate the weighted interconnection while keeping the circuitry simplicity. Note that the approximation is reasonable. When $x = 5v$, the output current I_o is the same as the function f_1. From Figure 4.2, f_1 is a good approximation of a linear function when w is within $[1.5, 3.5]$, and is saturated when w is outside of $[1.5, 3.5]$. This saturation feature can avoid the over-learning. As a consequence, the learning speed may be improved.

4.2.2 The Neuron Circuit

Since the output of the neuron is digital, the threshold function is used as the transfer function S_j in the neuron model. The neuron circuit implementation is

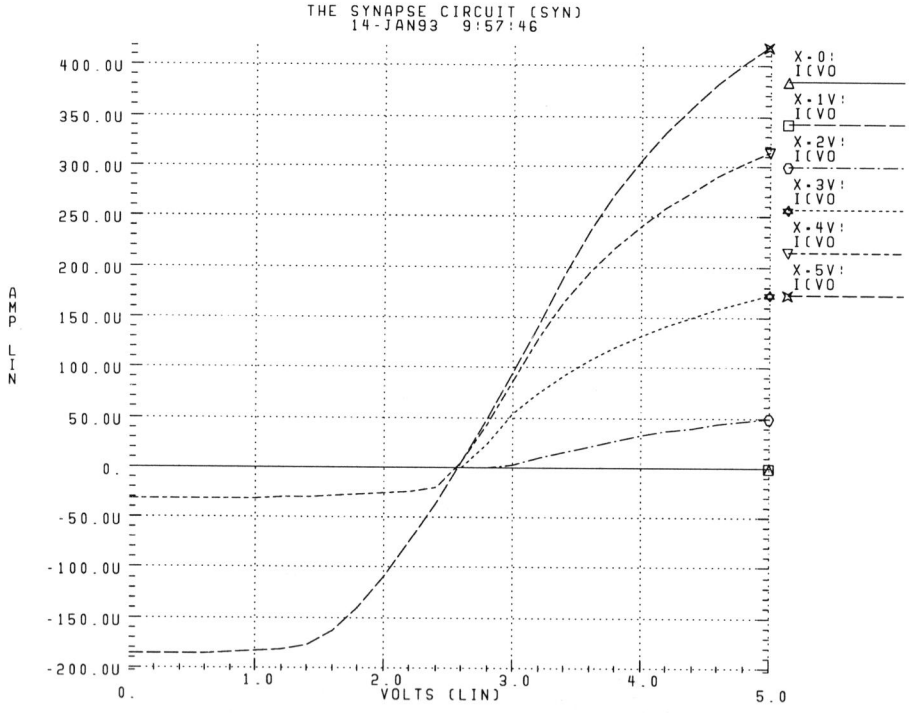

Figure 4.2 The synapse circuit characteristics

A Compact VLSI Implementation of Neural Networks

shown in Figure 4.3. Note that the input of the neuron circuit is the summation of current. The circuit consists of a current to voltage converter followed by an inverter. Transistor $M1$ and $M2$ construct a current to voltage converter. The reference voltage V_r is used to adjust the threshold point by changing the $M2$ current. Transistor $M3$ and $M4$ construct a simple inverter. This circuit characteristics simulated by HSPICE is shown in Figure 4.4.

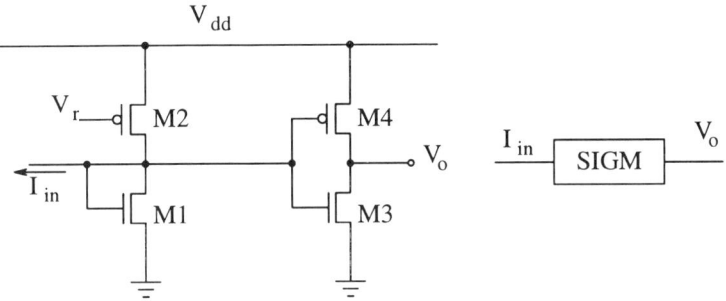

Figure 4.3 The neuron circuit

4.2.3 The Analog Current Switch Circuit

In the learning procedure the error computation e is needed. The error is

$$e = t - y, \qquad (4.8)$$

where t is the target output and y is the output of a neuron. When t and y are digital, the difference can be -1, 0 or 1. So we need two switches to analog this computation. The multiplication computation ew, during the learning procedure is implemented by the circuit shown in Figure 4.5. The transistor $M1$, $M2$ and $M7$ construct a synapse circuit, the output current I_{o1} is

$$\begin{aligned} I_{o1} &= kI_1 y \\ &= kf_1(w)y. \end{aligned} \qquad (4.9)$$

If the reference voltage V_{rn} is adjusted to make

$$I_{M5} = I_{M2}, \qquad (4.10)$$

the transistor $M3$, $M4$, $M5$ and $M6$ construct a current inversion. Since

$$I_{M1} = I_{M3} = I_{M4} = I_{M6}, \qquad (4.11)$$

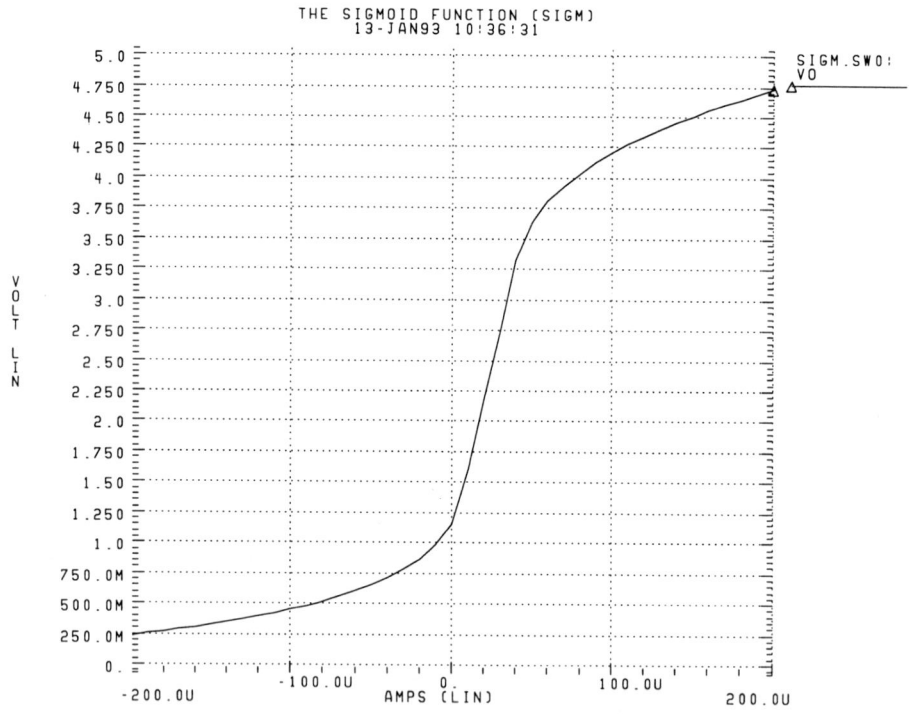

Figure 4.4 The neuron circuit characteristics

A Compact VLSI Implementation of Neural Networks

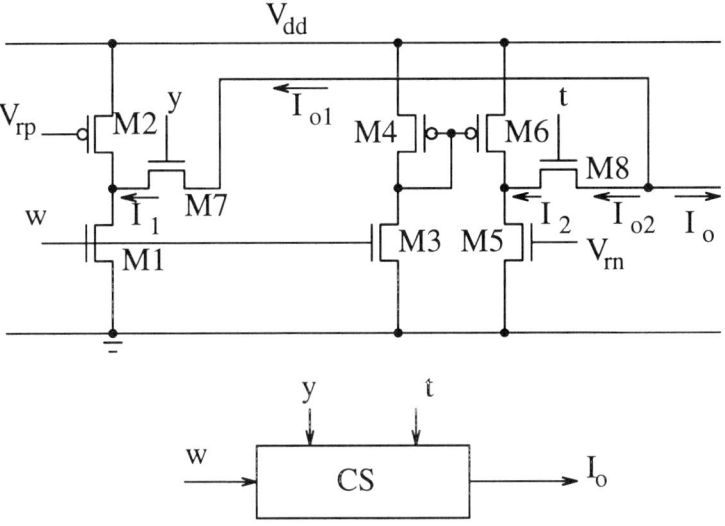

Figure 4.5 The analog current switch circuit

$$
\begin{aligned}
I_2 &= I_{M5} - I_{M6} \\
&= I_{M2} - I_{M1} \\
&= -I_1.
\end{aligned}
\quad (4.12)
$$

The circuit output current I_o is

$$
\begin{aligned}
I_o &= -(I_{o1} + I_{o2}) \\
&= -(kI_1 y + kI_2 t) \\
&= kf_1(w)(t - y) \\
&= kf_1(w)e.
\end{aligned}
\quad (4.13)
$$

4.2.4 The Error Generation Circuit

There are some problems when the target t and the neuron output y are used to control the positive and negative switches separately in Figure 4.5. Let us consider $t = y = 1$. The positive and negative currents may not be the same due to the parametric variation, which means $I_{o1} \neq I_{o2}$. This causes the output current to be nonzero, i.e., $I_o \neq 0$. This can be avoided by using d_+ and d_- to control positive and negative switches instead of t and y. The digital signal d_+ and d_- are defined as

$$d_+ = ty\prime, \qquad (4.14)$$

$$d_- = t\prime y. \qquad (4.15)$$

The logical circuit to generate d_+ and d_- is shown in Figure 4.6.

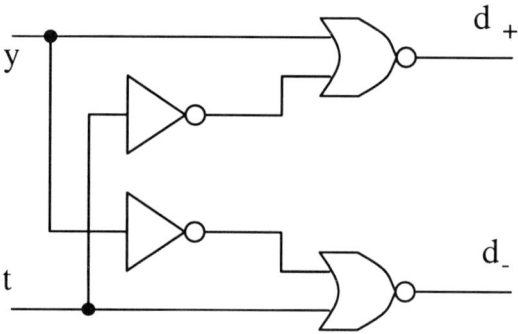

Figure 4.6 The modified error generation circuit

In this case the positive and negative switches can not be on at the same time. The positive and negative switches are off when $t = y = 1$, so there is no output current no matter what the positive and negative currents are. When d_+ and d_- replace t and y in Figure 4.5, the output current I_o is

$$I_o = kf_1(w)(d_+ - d_-) \qquad (4.16)$$

The output current I_o versus the error e is shown in Figure 4.7. Note that the output current in the modified circuit is zero for the region of $|e| < \delta$ region, not just at the origin point. So the stable state is not a point but a region. It helps the convergence of the learning procedure. This is another feature of the modified error function.

A Compact VLSI Implementation of Neural Networks

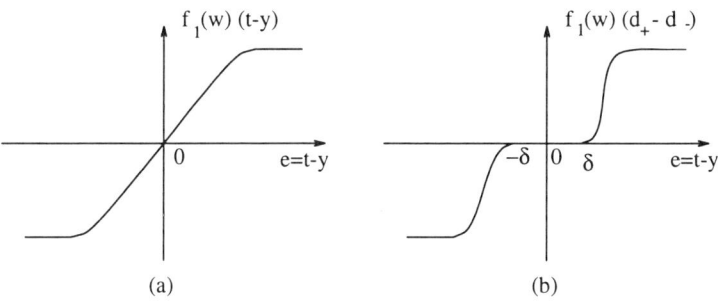

Figure 4.7 The output current versus the error e

4.2.5 The Current Copy Circuit

In neural computations, some variables need to be used many times. Variable copiers have to be implemented. The hard wire interconnection is a natural way to implement voltage copier. The current copier consists of two current inverters shown in Figure 4.8. The transistors $M1$–$M4$ construct first current

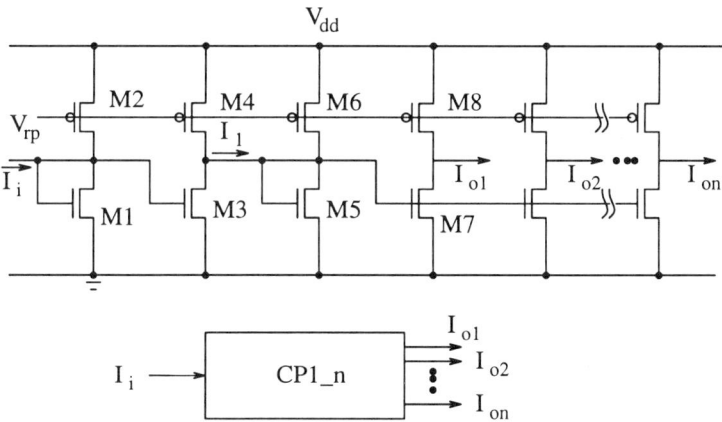

Figure 4.8 The current copier circuit

inverter. The transistors $M5$–$M8$ construct the other. The transistors $M1$ and $M3$, $M2$ and $M4$ form two current mirrors. Considering

$$I_{M1} = I_{M3}, \qquad (4.17)$$

$$I_{M2} = I_{M4}, \qquad (4.18)$$

the output current of the current inverter is

$$\begin{aligned} I_1 &= I_{M4} - I_{M3} \\ &= I_{M2} - I_{M1} \\ &= -I_i. \end{aligned} \quad (4.19)$$

So the output current of the current copier is

$$I_{o1} = I_{o2} = \cdots = I_{on} = -I_1 = I_i. \quad (4.20)$$

Note that the n-output current copier needs $2n + 6$ transistors, i.e., it needs to add two transistors for each additional output.

4.3 THE CIRCUIT IMPLEMENTATION EXAMPLE

4.3.1 A Feedforward Neural Network (FNN) Architecture with Learning

Consider the basic structure for a multilayer FNN. This process begins with the input (first) layer, and ends with the output (last) layer. Any layer between the input and output layers is often referred to as a hidden layer. This simple regular repetitive structure constitutes the FNN. The basic operation for each neuron in any layer may be described as

$$y_j = S_j(\sum_{i=1}^{N} w_{ji} x_i + \theta_j), \quad (4.21)$$

where y_j is the output of neuron j, $S_j( )$ is a nonlinear (sigmoid) function, w_{ji} is the synapse weight from the output of the neuron i to the input of the neuron j, and x_i is either the output of the previous layer neuron i or the network input i. θ_j is a bias.

The learning rule is important in ANN algorithms, and is also complicated to implement in hardware. The widely used error Backpropagation rule attempts to minimize the squared output error for training sample input-output pairs (or patterns). For every training sample, it modifies each weight in proportion to the partial derivative of the squared error with respect to that weight. For each desired target p, say $t_p^T = (t_{p1}, \ldots, t_{pn})$, the squared error function is defined as follows [7],

$$E_p = \frac{1}{2}\sum_{j=1}^{n}(t_{pj} - y_{pj})^2. \qquad (4.22)$$

The Backpropagation rule defines the change due to applying pattern p by

$$\triangle_p w_{ji} = -\eta \frac{\partial E_p}{\partial w_{ji}}, \qquad (4.23)$$

where $\eta > 0$ is the learning rate. Equation (4.23) has been modified for VLSI implementation [5]. This modified learning rule removes all the sigmoid derivatives which is complicated to implement by VLSI technology. The modified learning rule is defined as

$$\dot{w}_{ji} = \eta e_j y_i, \qquad (4.24)$$

where $\dot{w}_{ji}$ is the time rate of change of the weight w_{ji}, and e_j given by

(a) if neuron j is a member of the (last) output layer, then

$$e_j = t_j - y_j. \qquad (4.25)$$

(b) if neuron j is not a member of the output layer, then

$$e_j = \sum_k e_k w_{kj}, \qquad (4.26)$$

where k is the index for neuron units in the next layer to which neuron j is connected.

Note that Equation (4.24) is different from the update rule in [5]. The weight w_{ji} is dynamically and continuously updated. It is changed during each pattern presented. So the summation is removed from Equation (4.24).

4.3.2 Modified FNN Architecture

An FNN architecture with learning, discussed in the previous section, is modified based on the designed building block circuits. The feedforward computation is revised as

$$y_j = S_j(\sum_{i=1}^{N} k f_1(w_{ji})x_i + f_1(\theta_j)), \qquad (4.27)$$

where θ_j can be treated as a weight interconnection to neuron j from a neuron whose output is always one.

Table 4.1 ENCODER TEST RESULTS

Problem	Trials	Max	Min	Average	S.D.
4-2-4	20	114	5	25	5

The learning rule is similar to the modified learning rule described in Section 4.3.1. But the difference computation $t_j - y_j$ is replaced by $d_{+j} - d_{-j}$. So the update law can be rewritten as

(a) if neuron j is a member of the output layer,

$$\dot{w}_{ji} = \eta(d_{+j} - d_{-j})y_i; \qquad (4.28)$$

(b) if neuron j is a member of the hidden layer which connects to output layer,

$$\dot{w}_{ji} = \eta(\sum_k (d_{+k} - d_{-k})\, w_{kj})\, y_i. \qquad (4.29)$$

The modified FNN architecture is tested at the function level by applying it to solve the 4-2-4 encoders [8]. The test results are listed in Table 4.1. We have run 20 trials with the weights initialized to the different random values in each case. The choice of initial weights might have a great influence on the learning time required. From Table 4.1, the longest trial required 114 epochs (one epoch means that all training patterns are presented once at the network), the shortest required 5 epochs, the average over all the runs was 25 epochs, and the standard deviation was 5 epochs.

4.3.3 Circuit Simulation

The block diagram of the circuit implementation for this modified FNN architecture is shown in Figure 4.9. The modified FNN consists of three layers: the input layer, the hidden layer, and the output layer. The input layer is simply two voltage nodes of the external input denoted by x_1 and x_2. The hidden layer has one neuron with output denoted by y_{h1} and bias denoted by θ_{h1}. w_{hji} is the interconnection between the ith input of input layer and the jth neuron of the hidden layer (here $j = 1$). The output layer has two neuron with output denoted by y_{o1} and y_{o2}, and bias denoted by θ_{o1} and θ_{o2}. w_{okj} is the interconnection between the jth neuron of the hidden layer and the kth neuron of the output layer.

A Compact VLSI Implementation of Neural Networks

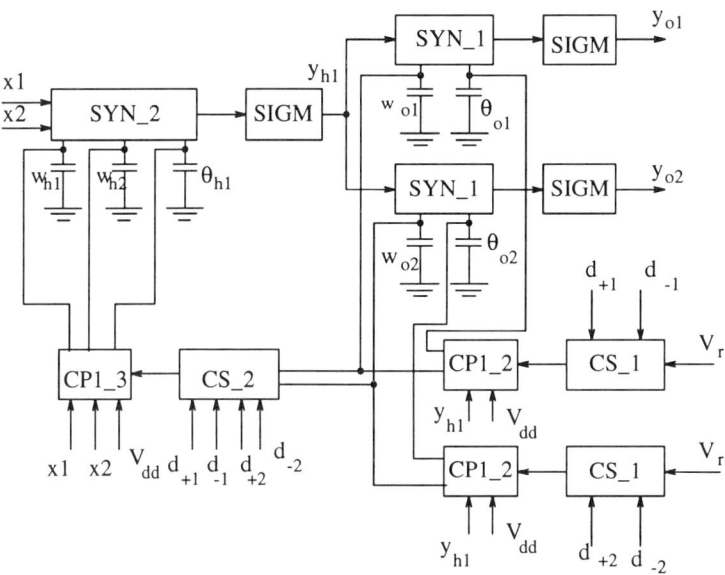

Figure 4.9 Block diagram of the 2-1-2 FNN circuit

Figure 4.9 consists of two components, the feedforward neural network subcircuit and the learning subcircuit. The block SIGM is a threshold function and is implemented in Figure 4.3. The block SYN_1 is a weighted interconnection and is implemented in Figure 4.1, Block SYN_2 consists of two block SYN_1. The block CS_1 is a analog current switch and is implemented in Figure 4.5. Block CS_2 consists of two block CS_1. Block CP1_n is a n-output current copier shown in Figure 4.8, and followed by n switches.

We use this network to learn the 2-1-2 encoders task. The circuit simulation by HSPICE is shown in Figure 4.10. The waveform H is the output voltage of the hidden neuron node. The $T0$ and $T1$ are the voltage of target nodes. The $Y0$ and $Y1$ are the output voltage of the output neuron nodes. Note that the circuit converges after 200ns. The simulation results show that the network successfully learns the 2-1-2 encoders problem.

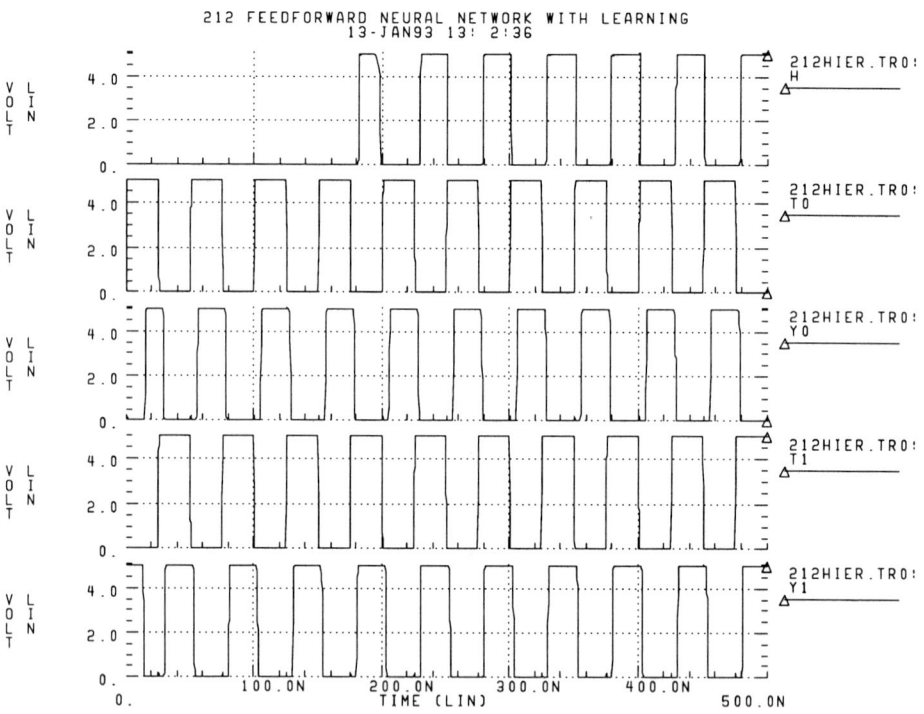

Figure 4.10 HSPICE simulation results of the 2-1-2 FNN circuit with learning

4.4 EXPANDING THE NETWORK WITH MULTIPLE CHIPS

The large scale network which cannot be expressed by a single chip needs the expanding by interconnecting multiple chips. Two classes of building blocks discussed in the previous sections are suitable for constructing the module chips which can expand the FNNs. One module chip is designed to be used as any layer. The block diagram of the architecture of a general $n \times m$ module chip is shown in Figure 4.11, where n is the number of inputs and m is the number of outputs. There are four I/O signals, x, y, e_p and e_n. x is a n-dimensional input vector, y is a m-dimensional output vector, e_p is a m-dimensional error vector from the previous higher layer, e_n is a n-dimensional backpropagated error vector to the next lower layer. The block func_n(xm) in Figure 4.11 simply consists of n(xm) blocks of func which can be found in the previous section.

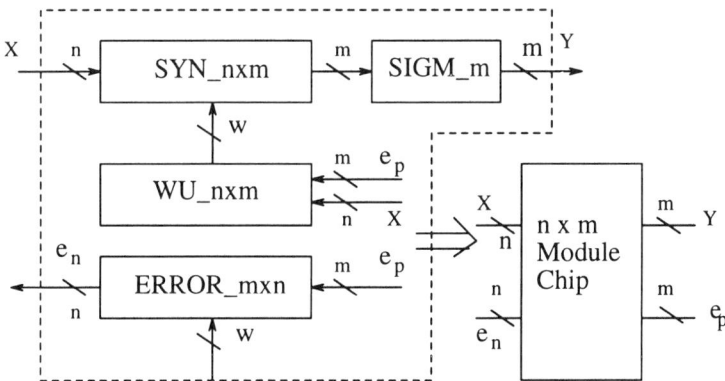

Figure 4.11 The $n \times m$ module chip architecture

Figure 4.12 illustrates a simple example of how this module chip can be used to expand vertically and horizontally. If $n = km$, then a $n \times km \times m$ fully connected two layer FNN with learning can be implemented using $(k+1)\, n \times m$ module chips [9].

The input layer is composed of just the input voltage nodes x. The module chips in the middle form a hidden layer. The outputs of the hidden layer are applied to the inputs of the output layer. In the output layer, the desired target vector is supplied through the error vector e_p. If there is no lower layer, then the error vector e_n is not propagated backward any further. Otherwise, it is propagated through the error vector e_p of the module chip in the lower layer.

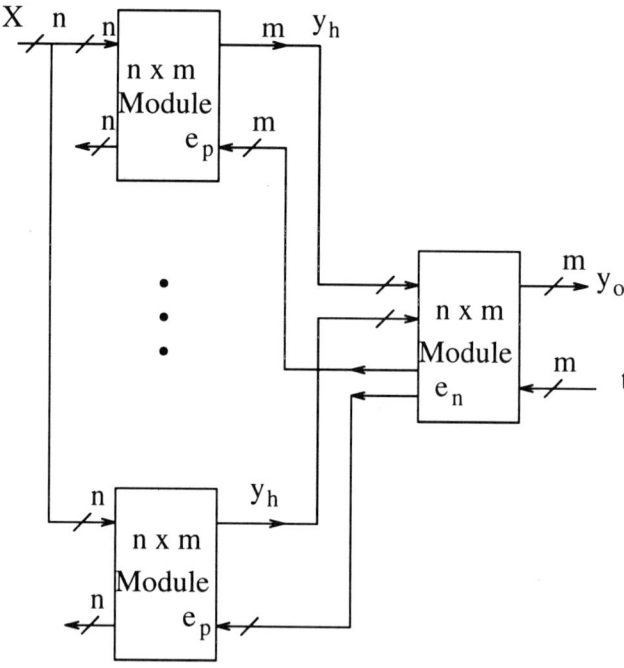

Figure 4.12 A $n \times km \times m$ two layer FNN architecture with learning

4.5 CONCLUSIONS

In this chapter, we have studied how the circuit implementations can efficiently approximate the neural network operations and designed a class of compact circuit building blocks. A modified feedforward neural network architecture, with learning, has been implemented by the designed circuits. This architecture which is applicable to encoder problems has been tested in the functional and circuit levels. Some applications need large scale networks. The module chip design and its use to expand the scale of networks by interconnecting multiple chips are discussed.

REFERENCES

[1] J. Alspector, "VLSI Architectures for Neural Networks," in *Neural Networks: Concepts, Applications, and Implementations*, P. Antognetti & V. Milutinovic, eds. #I, Prentice Hall, 1991, 180–213.

[2] E.A. Vittoz, "Future of Analog in the VLSI Environment," in *IEEE International Symposium on Circuits and Systems*, 1990, 1373–1375.

[3] C. Schneider & H. Card, "CMOS Implementation of Analog Hebbian Synaptic Learning Circuits," in *International Joint Conference on Neuron Networks* #I, 1991, 437–442.

[4] Y. Maeda, H. Yamashita & Y. Kanata, "Learning Rules for Multilayer Neural Networks Using a Difference Approximation," in *International Joint Conference on Neuron Networks*, 1991, 628–633.

[5] Fathi M.A. Salam & M. Choi, "An All-MOS Analog Feedforward Neural Circuit with Learning," in *IEEE International Symposium on Circuits and Systems*, 1990, 2508–2511.

[6] D. Brown & A. Scott, *Design Rules and Process Parameters for the Northern Telecom CMOS4s Process*, Report IC90-01, Canadian Microelectronics Corporation, Feb.,1990.

[7] D.E. Rumelhart & J.L. McClelland, *Parallel Distributed Processing: Explorations in the Microstructure of Cognition*, MIT Press, Cambridge, MA, 1986.

[8] S. E. Fahlman, *An Empirical Study of Learning Speed in Back-Propagation Networks*, Technical Report CMU-CS-88-162, Carnegie-Mellon University, Sept., 1988.

[9] M. Choi & F. M. A. Salam, "Implementation of Feedforward Artificial Neural Nets with Learning Using CMOS VLSI Technology," in *IEEE International Symposium on Circuits and Systems*, 1991, 1509–1512.

5
AN ALL-DIGITAL VLSI ANN

Brian White and Mohamed I. Elmasry

5.1 INTRODUCTION

In recent years, there has been a great deal of research activity in artificial neural networks (ANN) in the area of simulation and hardware implementations [1-6]. Because of the special features offered by ANNs, such as the capability to learn from examples, adaptation, parallelism, fault tolerance and noise resistance, they have been applied to a number of real-world problems including image and speech processing [3,6-9]. To enhance the impact of ANNs and broaden the area of applications, it is imperative that ANNs benefit from the state-of-the-art VLSI and ULSI implementation technologies. Because these technologies are basically a digital implementation medium, ANNs must be adapted to an all-digital implementation approach. To illustrate this thesis, the research work reported in this paper offers a practical example of adapting an ANN model to an all-digital VLSI implementation.

An all-digital ANN VLSI implementation offers several advantages over its analog counterpart.

1. In most real-world applications, ANNs are embedded in existing *digital* hardware/software systems [10]. An all-digital ANN VLSI implementation solves the compatibility problem.

2. Real-world applications require *large* neural networks of more than 10,000 neurons and synapses [8]. Digital VLSI/ULSI is more appropriate at this level of complexity, whereas analog VLSI/ULSI suffers from noise susceptibility and difficulties in fabricating high precision resistors and capacitors.

3. Larger ANNs are most likely implemented in a multi-chip set and the analog implementation would make it more difficult to transfer signals from chip to chip, and also to match board-level capacitive loads and time constants [11].

4. At any given time, digital VLSI technology is always more mature than its analog counterpart, in terms of fabrication technology and simulation and design automation tools. It also offers a wide range of fabrication technologies, including such technologies as Field-Programmable Gate Arrays [12] for rapid prototyping.

5. Real-world examples may suffer from I/O bottlenecks [9], which are best addressed by digital techniques such as input buffers, shift registers, and pipelining. Moreover, power dissipation reduction techniques, such as dynamic logic and complementary operation, can be used.

6. Digital implementation offers a homogeneous implementation environment between the processing elements and the on-chip or external memory storage.

Other advantages of digital VLSI implementation apply specifically to the Neocognitron ANN [13-15] that is considered here. The Neocognitron has a locally connected network with *repetitive* use of the *same* weights; this fits well with time division multiplexing of digital hardware [8,9]. For *temporally sparse* (only a small subset of inputs actually take part in the computation) ANNs such as the Neocognitron (as shown in Section 5.4), fully parallel analog multipliers look less attractive and a digital shared resource approach is again more suitable [16]. The Neocognitron model also uses a variety of complex functions, not just a single activation function. It is easier to implement these functions digitally in lookup tables, than to attempt design of a number of appropriate analog elements.

There are also some shortcomings of digital VLSI that must be resolved in order to implement ANNs. Most ANN neuron calculations involve a weighted sum of the neuron inputs. The multiplier required for this multiply-accumulate operation is slow and consumes large area in a digital VLSI implementation. However, if the neural network model is carefully modified to approximate all multiplying factors by powers of 2 (to replace a multiplier by a shifter), this problem can be eliminated, as shown in Section 5.3, and an all-digital implementation becomes very attractive.

It is also important to indicate why the Neocognitron neural network model, possibly the most complicated network ever developed [3], was investigated for all-digital VLSI implementation in this work, rather than a simpler model. Our approach was to select a real-world application area (optical character recognition) and let the application dictate the selection of the ANN. Realistic two-dimensional image recognition systems require position, scale and rotation invariance, as well as tolerance of noise and distortions. The Neocognitron uses a hierarchical type of network to produce outputs invariant to position, size and small distortions, and is capable of complex character recognition [3,6]. Its architecture of neurons organized in two-dimensional planes, local interconnections and repeated weights also matches the general features required [8,9] for successful optical character recognition by ANNs. Section 5.2 describes the Neocognitron model in detail.

This paper proposes a new neural network model called the Digi-Neocognitron (DNC), described in Section 5.3, which has the same pattern recognition performance as the Neocognitron (NC) but is better suited for digital VLSI implementation. The DNC model is derived from the NC model by a combination of preprocessing approximations and definition of new model functions. Multiplication and division are eliminated by conversion of factors to powers of 2, requiring only shifts. Bit widths are restricted, and complex functions are implemented with lookup tables, carefully tailored to retain performance. A methodology for conversion of a particular NC model to a DNC model is described; this procedure utilizes the Neocognitron learning capability and fault tolerance to compensate for approximations. Since the learning procedure uses the NC model, it is performed off-line.

Section 5.4 reports on the simulation results of a character recognition example, which proves the feasibility of this approach. The NC model is trained on centered versions of the input patterns, which are alphanumeric characters in a 12 by 12 pixel array [17], and the conversion methodology is followed to derive a DNC model. Both neural networks are tested on translations of the original patterns with varying levels of added random noise; this quantifies recognition performance, translation invariance and noise tolerance. Both ANNs have very good recognition rates in the presence of large amounts of noise, and the DNC model has no deterioration from NC performance. Both ANNs can also classify input patterns as unknown, which significantly reduces the undetectable error rate (wrong classifications); this is very important for commercial applications [18].

Section 5.5 reports on the advantages of the DNC model as compared to the NC model for VLSI implementation. The area-delay product for implementation of the multiplication in the weighted sum calculation is improved by 2 to 3 orders of magnitude. I/O and memory requirements are substantially reduced by representation of weights with 3 bits or less and neuron outputs with 4 bits or 7 bits.

5.2 NEOCOGNITRON NEURAL NETWORK MODEL

The Neocognitron, proposed by Fukushima [13-15], is an artificial neural network model for visual pattern recognition. It has generated interest due to its capabilities for shift-invariant and deformation-resistant pattern recognition, and has been applied to character recognition tasks [3,14,19-21]. The Neocognitron has some similarities to the human visual system, and is self-organizing i.e it performs unsupervised learning, or learning without a teacher. We focus our attention, as in [22], on the more popular unsupervised learning model, although a supervised learning version also exists [19].

The basic architecture of the Neocognitron neural network model is shown in Figure 5.1. There are four different kinds of neurons, called S-cells, C-cells, Vs-cells and Vc-cells. All of the cells have non-negative outputs. The Vc-cells and Vs-cells provide inhibitory inputs to suppress outputs of S-cells and C-cells, respectively. The Neocognitron is a multi-layer neural network, with two-dimensional input layer connected to a cascade of layers consisting of planes of S-cells and planes of C-cells. S-cells can be interpreted as feature detectors, and C-cells tolerate position errors by averaging S-cell outputs in a small area. Only interconnections to S-cells have modifiable weights, determined during learning; other interconnection weights are fixed, but are adjustable architectural parameters that influence learning and recognition performance. Typically, each S-cell plane develops weights that look for a different type of feature, and the number of planes can be different for each level. There is spatially local interconnect, with interconnection weights to a given neuron defined only in a small connection area, indicated by the circles in Figure 5.1. These weights are also constrained to be translationally invariant, with the same set of weights applying at all two-dimensional positions within a cell plane. These features give the Neocognitron capabilities for toleration of translations and deformations. At higher levels, cells respond to more complicated features, and the effective receptive field becomes larger. The number of neurons in a cell-plane reduces at higher levels, until at the final (recognition) level there is only one neuron per plane, each representing a recognition class.

The Neocognitron neural network model can be described mathematically as follows [14] . The S-cell output is

An All-Digital VLSI ANN

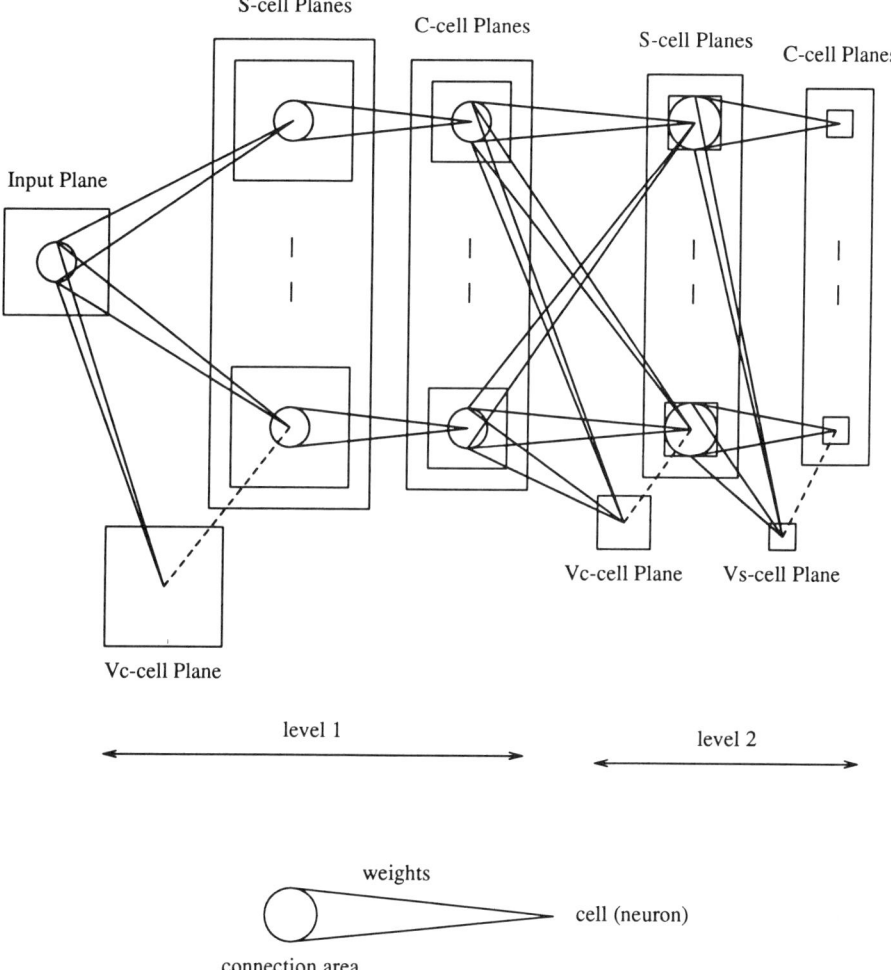

Figure 5.1 Architecture of the neocognitron neural network

$$u_{Sl}(k,\underline{n}) = r_l \phi \left[\frac{1 + \sum_{\kappa=1}^{K_{C\,l-1}} \sum_{\underline{\nu} \in A_l} a_l(\kappa, \underline{\nu}, k)\, u_{Cl-1}(\kappa, \underline{n}+\underline{\nu})}{1 + \frac{r_l}{1+r_l} b_l(k)\, v_{Cl}(\underline{n})} - 1 \right] \quad (5.1)$$

where the nonlinear activation function is

$$\phi[x] = max(x, 0). \quad (5.2)$$

Here $u_{Sl}(k, \underline{n})$ describes the output of an S-cell at two-dimensional coordinate $\underline{n}$ in the k-th plane in level l. The excitation weighted sum in the numerator sums the modifiable weight a_l multiplied by the previous level C-cell outputs, but only over the small connection area A_l. The sum is over all $K_{C\,l-1}$ planes in the previous level; this degenerates to the single input layer plane for level 1 S-cells (see Figure 5.1). The a_l weights are developed in the learning process; note that they are not a function of position $\underline{n}$. The inhibitory term in the denominator has a non-linear effect on the output, and involves the modifiable weight b_l developed during learning (represented by the dotted line in Figure 5.1), the Vc-cell output and the selectivity parameter r_l. The selectivity is very important in the learning process; larger r_l cause S-cells to be more selective in their response, due to more inhibition, but also less tolerant of distortions. The S-cell output is non-zero only if the excitation sum is greater than the inhibitory term; the inhibitory cells suppress output when irrelevant features are presented.

The inhibitory Vc-cell output is given by a weighted RMS average of the previous level C-cell outputs over the same connection area A_l:

$$v_{Cl}(\underline{n}) = \left[\frac{1}{csum} \sum_{\kappa=1}^{K_{cl-1}} \sum_{\underline{\nu} \in A_l} c_l(\underline{\nu})\, u_{Cl-1}^2(\kappa, \underline{n}+\underline{\nu}) \right]^{1/2} \quad (5.3)$$

The c_l are fixed weights, which are usually [23] taken as a two-dimensional Gaussian with adjustable variance σ_c. Here we have changed the normalization (from Fukushima) such that

$$c_l(0) = 1 \quad (5.4)$$

An All-Digital VLSI ANN

and this then requires division by the sum

$$csum = \sum_{\kappa=1}^{K_{Cl-1}} \sum_{\underline{\nu} \, \epsilon \, \dot{A}_l} c_l(\underline{\nu}) \qquad (5.5)$$

to obtain the same result for $v_{Cl}(\underline{n})$. This normalization process does not change the Vc-cell and S-cell outputs, and is not necessary for the Neocognitron. It is necessary for the Digi-Neocognitron model, as it provides a consistent range of values for the c_l fixed weights, which aids the approximations discussed in Section 5.3.

Interconnections from the S-cells to the C-cells have fixed weights, which are adjustable in the model architecture but not modified by learning. The C-cell output is given by

$$u_{Cl}(k, \underline{n}) = \psi \left[\frac{1 + \sum_{\underline{\nu} \, \epsilon \, D_l} d_l(\underline{\nu}) \, u_{Sl}(k, \underline{n} + \underline{\nu})}{1 + v_{Sl}(\underline{n})} - 1 \right] \qquad (5.6)$$

where the saturation function is

$$\psi[x] = \frac{x}{\alpha_l + x}, \quad x \geq 0 \qquad (5.7)$$
$$= 0, \quad x > 0$$

with saturation parameter α_l. This function limits response to the interval [0,1]. The excitation summation is over a single small connection area D_l, with the previous level S-cell outputs (for the same plane only) multiplied by d_l fixed weights, which are taken as a two-dimensional Gaussian with variance σ_d. Optionally, there may exist inhibitory Vs-cells in the architecture, which provide a form of lateral inhibition, with non-zero C-cell output only if the excitation sum is greater than the inhibitory term. The Vs-cell output is given by :

$$v_{Sl}(\underline{n}) = \frac{1}{K_{Sl}} \sum_{\kappa=1}^{K_{Sl}} \sum_{\underline{\nu} \, \epsilon \, D_l} d_l(\underline{\nu}) \, u_{Sl}(\kappa, \underline{n} + \underline{\nu}) \qquad (5.8)$$

This is a weighted average of activity in the connection area, but now taken across all K_{Sl} preceding level S-cell planes.

The unsupervised learning rules for the Neocognitron will not be covered in much detail, except to indicate the modification required due to our change of c_l normalization. In the learning process, *representative* S-cells with the strongest output in columns (centered on connection areas) through the S-cell planes are selected, at most one per S-cell plane. Interconnection weights over the connection area are reinforced in proportion to the input activities, and then replicated to all positions in the plane.

The rules as originally stated by Fukushima are, for *representative* $u_{Sl}(K, \underline{N})$,

$$\Delta a_l(\kappa, \underline{\nu}, K) = q_l \, c_l(\underline{\nu}) \, u_{Cl-1}(\kappa, \underline{N}+\underline{\nu}) \tag{5.9}$$

$$\Delta b_l(K) = q_l \, v_{Cl}(\underline{N}) \tag{5.10}$$

where q_l is a reinforcement parameter which affects the speed of weight development and type of feature detectors that emerge. With the renormalization to $c_l(0) = 1$, the first learning rule becomes

$$\Delta a_l(\kappa, \underline{\nu}, K) = q_l \, \frac{c_l(\underline{\nu})}{csum} \, u_{Cl-1}(\kappa, \underline{N}+\underline{\nu}). \tag{5.11}$$

As discussed above, this normalization change is not necessary for the Neocognitron model and yields the same results as the Neocognitron. It is necessary for the Digi-Neocognitron model.

5.3 DIGI-NEOCOGNITRON (DNC) : A DIGITAL NEURAL NETWORK MODEL FOR VLSI

5.3.1 Approach to VLSI Implementation

The need for VLSI implementation is considered at the highest level of abstraction possible, namely the neural network model itself. The Neocognitron (NC) neural network model functions and weight values are changed to facilitate a digital VLSI circuit implementation, and this results in a new neural network which we call the Digi-Neocognitron (DNC).

Multiplications and divisions are replaced by simple shifts by conversion of multiplying or dividing factors to powers of 2; shifters have a simple hardware implementation which substantially reduces the silicon area and propagation delays as compared to multipliers. Bit ranges are restricted as much as possible. Complex functions are replaced by lookup tables, which can be implemented by simple combinatorial logic or memory arrays, due to the reduced bit ranges.

Bit range restrictions in the DNC model start from decisions (tested by simulation) on the representation of the different types of cell outputs. Vc-cell and C-cell outputs are in the interval [0,1] and are reasonably represented with 4 b (4 binary places to the right of the binary point). There is no explicit range constraint on Neocognitron S-cell outputs, but in principle the output is in the interval [0,1] as discussed by Fukushima [15]. However, the DNC model approximation of modifiable weights as powers of 2 (discussed below) can result in S-cell outputs greater than 1 if excitatory weights are rounded up and inhibitory weights rounded down. In our simulations the DNC S-cell outputs have been less than 2, but a representation of 7 b (with 4 binary places) is used to provide a practical safety margin (represent outputs up to 8).

The following sections describe the DNC neural network model, including preprocessing approximations that would always be done in a DNC simulator, and model functions that would be implemented in VLSI hardware. The model is described functionally, including necessary bit widths; for more details see [24]. Detailed VLSI architectural issues are addressed in [25].

5.3.2 Power-of-2 Representations

In digital circuit implementations, the multiplication (or division) of binary numbers requires large chip area and causes speed deterioration; a method for reducing this hardware complexity approximates binary numbers by the sum of a limited number of signed power-of-two terms [26,27]. Multiplication or division by a power of 2 is a shift operation, with a simple hardware implementation, which reduces the silicon area and propagation delays, as shown in Section 5.5. Note that a power of 2 in the range 2^{-m} to 2^n has only $(n+m+1)$ distinct values, so the number of bits required for input is also reduced considerably.

The DNC model uses single-term power-of-2 approximations in all cases except one, which requires a two-term expansion. The description of the DNC model uses the following notation, similar to [27] but modified for the case of positive numbers and fractional parts.

Denote a particular single-term power-of-two space (set of admissible values in the representation) as

$$P_{m,n} = \{2^m, 2^{m+1}, ..., 2^{n-1}, 2^n\}$$

where m and n are integers, with $m \leq n$. Define

$$\langle x \rangle_{2^m, 2^n \,; \, b} \tag{5.12}$$

to denote rounding x to the nearest term belonging to the representation $P_{m,n}$, with bias $b \in [0, 1]$ to rounding up. If x is outside the range of the representation, the appropriate endpoint of the representation is used as the approximation.

Bias 0.0 selects the power of 2 that is less than or equal to the binary number, bias 1.0 selects the power of 2 that is greater than or equal to the number, and bias 0.5 selects the closest power of 2. We can simply denote these as

$$\langle x \rangle_{2^m, 2^n \,; \, le} \doteq \langle x \rangle_{2^m, 2^n \,; \, 0.0}$$

$$\langle x \rangle_{2^m, 2^n \,; \, ge} \doteq \langle x \rangle_{2^m, 2^n \,; \, 1.0}$$
$$\langle x \rangle_{2^m, 2^n} \doteq \langle x \rangle_{2^m, 2^n \,; \, 0.5} \tag{5.13}$$

Extending this concept to the one case where two terms are required, denote a particular two-term power-of-two space (set of admissible values in the representation) as

$$P^2_{m,n} = \{\, y \mid y = 2^p + s2^k,\ y > 0 \,\}$$

where m, n are integers, with $m \leq n$, $p, k \in \{m, m+1, ..., n-1, n\}$ and $s \in \{-1, 0, 1\}$.

Define

$$\langle\langle x \rangle\rangle_{2^m, 2^n} \tag{5.14}$$

to denote rounding x to the nearest term belonging to the representation $P^2_{m,n}$.

5.3.3 Digi-Neocognitron : Inhibitory Vc-cells

5.3.3.1 *Preprocessing*

Equation 5.3 describes the inhibitory Vc-cell output function of the NC model. The normalization change of Equation (5.4) provides a consistent range of values for c_l that allows approximation by values $\bar{c}_l$ restricted to 0 and simple powers of 2, as shown in Table 5.1.

Table 5.1 Preprocessing of Fixed Weights

Fixed Weight c_l, d_l Interval	Fixed Weight $\bar{c}_l, \bar{d}_l$ Approximation	Interpretation
[0.0 , 0.1)	0	force 0
[0.1 , 0.4)	1/4	shift 2 R
[0.4 , 0.75)	1/2	shift 1 R
[0.75 , 1.0]	1	shift 0

The normalization sum of the fixed weights in Equation (5.5) is now

$$C_\Sigma = \sum_{\kappa=1}^{K_C \, l-1} \sum_{\nu \in A_l} \overline{c_l(\nu)} \qquad (5.15)$$

To replace the division in Equation (5.3) by a shift operation, a power-of-2 approximation of Equation (5.15) is done as follows, for level l:

$$\overline{C_\Sigma} = \langle C_\Sigma \rangle_{1,\, 2048\,;\, le} \quad , \quad l = 1 \qquad (5.16)$$

$$= \langle C_\Sigma \rangle_{1,\, 2048\,;\, ge} \quad , \quad l > 1$$

The representation accommodates architectures having c_l mask size (connection area A_l) up to 7 by 7 and number of preceding planes K_{Cl-1} up to 128. At level $l = 1$, le rounding is used to increase the inhibit output to suppress noise in the S-cell plane outputs; at higher levels, ge rounding is used to decrease the inhibit (noise is already suppressed by level 1) to allow S-cells to be activated more easily.

5.3.3.2 DNC Model Functions

With the preprocessing approximations, the DNC model Vc-cell output is

$$\overline{v_{Cl}(\underline{n})} = \left[\frac{1}{\overline{C_\Sigma}} \sum_{\kappa=1}^{K_{cl-1}} \sum_{\underline{\nu} \in A_l} \overline{c_l(\underline{\nu})} \, \overline{u_{Cl-1}(\kappa, \underline{n} + \underline{\nu})}^2 \right]^{1/2}. \quad (5.17)$$

which is implemented as shown in Figure 5.2. The inputs $\overline{u_{Cl-1}}$ are lower level C-cell outputs or the input layer pattern, represented with 4b; on the input layer a 0 pixel is input as 0.0000 binary and a 1 pixel as 0.1111 binary for consistency of hardware implementation. Multiplication by $\overline{c_l}$ is implemented by forcing 0 or shifting 0-2 right, as per Table 5.1; results are truncated to 4b. To implement the squaring of $\overline{u_C}$ without a multiplier, a power-of-2 approximation of $\overline{u_C}$ is used to control shifting. This is defined in Table 5.2, and can be implemented in digital hardware with a simple combinatorial circuit. Results of the shift are truncated to 4b. Note that the *output* of the shifter is not restricted to powers of 2.

Summation over connection areas in Equation (5.17) requires a 16-b accumulator (adder plus register). Note that we are describing the DNC model functions not the VLSI architectural implementation, which could have parallel shifters feeding into pipelined carry-save adders, for instance. Division by $\overline{C_\Sigma}$ is implemented with a shift. The square root in Equation (5.17) is approximated with a lookup table as shown in Table 5.3; this can be implemented with simple combinatorial logic. The first four entries increase the inhibit at low activity levels for S-cell output noise suppression and to compensate for truncations. The final $\overline{v_C}$ output is 4 b.

Note that a shift operation is used to implement the squaring of $\overline{u_C}$, rather than a lookup table as used for the square root. The square root calculation is only done once per $\overline{v_C}$ calculation, whereas the $\overline{u_C}$ squaring is done for every term in the summation. In the VLSI implementation, it is likely that many summation terms will be calculated in parallel in identical pipelined systolic

An All-Digital VLSI ANN

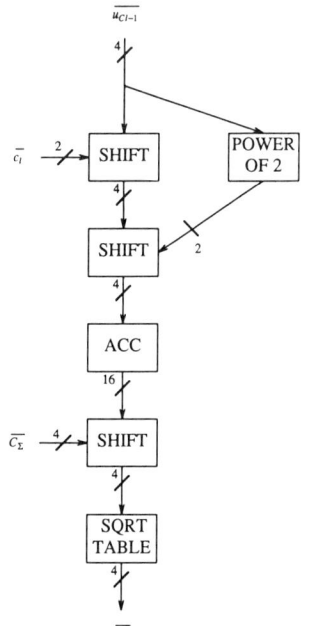
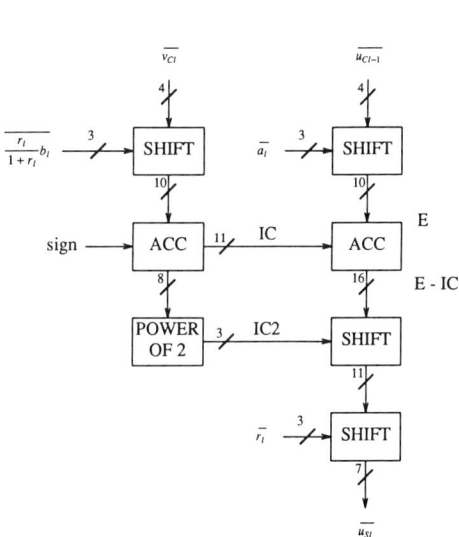

Figure 5.2 Digi-neocognitron Vc-cell block diagram, with bit widths indicated

Figure 5.3 Digi-neocognitron S-cell block diagram, with bit widths indicated

Table 5.2 Power of 2 Approximation for Squaring C-Cell Output

Input $\overline{u_C}$ (binary)	POWER OF 2 ($\overline{u_C}$) SHIFT Control
0000	Force 0
0001	Force 0
0010	Force 0
0011	Shift 2 R
0100	Shift 2 R
0101	Shift 1 R
0110	Shift 1 R
0111	Shift 1 R
1000	Shift 1 R
1001	Shift 0
1010	Shift 0
1011	Shift 0
1100	Shift 0
1101	Shift 0
1110	Shift 0
1111	Shift 0

Table 5.3 Square Root Table for Vc-Cell Output

Input (decimal)	Input (binary)	Output (binary)	Output (decimal)
0.0	0.0000	0.0100	0.25
0.0625	0.0001	0.0110	0.375
0.125	0.0010	0.0111	0.4375
0.1875	0.0011	0.1000	0.5
0.250	0.0100	0.1000	0.5
0.3125	0.0101	0.1001	0.5625
0.375	0.0110	0.1010	0.625
0.4375	0.0111	0.1011	0.6875
0.5	0.1000	0.1011	0.6875
0.5625	0.1001	0.1100	0.75
0.625	0.1010	0.1101	0.8125
0.6875	0.1011	0.1101	0.8125
0.75	0.1100	0.1110	0.875
0.8125	0.1101	0.1110	0.875
0.875	0.1110	0.1111	0.9375
0.9375	0.1111	0.1111	0.9375

cells, so it is important to reduce the area and delay of the squaring circuit. In this model, as shown in Figure 5.2, only the shifter delay appears on the critical path, with shift control determined in parallel. This achieves a critical path delay significantly less than a lookup table approach, and also uses less area. The accuracy of this shift implementation is the same as a lookup table for inputs up to and including 0.1000 binary. For larger inputs the output is greater than for an "accurate" lookup table, but it was found in simulation that this provided better results for $\overline{v_C}$ when combined with other truncations and approximations in the DNC model.

5.3.4 Digi-Neocognitron: S-cells

5.3.4.1 Preprocessing

Equation 5.1 describes the S-cell output function of the Neocognitron. Modifiable weights a_l and b_l (actually $r_l b_l / (1 + r_l)$) are approximated by powers of 2 as described below. For fine-tuning, all the a_l and b_l weights can be scaled by a factor f_l to improve alignment with the power-of-2 representation; this is specified on an input file to the DNC model simulator, as shown in the Appendix. For f_l close to 1.0 this is justifiable since a *small* change in all weight values is equivalent to using a different value of reinforcement q_l in the Neocognitron learning Equations 5.10 and 5.11.

Our experience with real examples is that the critical factor in emulating the Neocognitron with the Digi-Neocognitron is the inhibition term in the S-cells i. e. the Vc-cell output and the $r_l b_l / (1 + r_l)$ weight factor; this agrees with the conclusions of a fault tolerance analysis of the Neocognitron [28]. A power-of-2 approximation of $r_l b_l / (1 + r_l)$ consequently uses one or *two* power-of-two terms (as specified on the input file to the DNC model simulator shown in the Appendix):

$$\overline{r_l b_l(k)/(1 + r_l)} = \langle f_l\, r_l b_l(k)/(1 + r_l) \rangle_{1,\,64} \quad (5.18)$$

or

$$\overline{r_l b_l(k)/(1 + r_l)} = \langle\langle f_l\, r_l b_l(k)/(1 + r_l) \rangle\rangle_{1,\,64}. \quad (5.19)$$

One term is sufficient on lower levels but two terms are needed on higher levels.

A power-of-2 approximation of the modifiable weight a_l, after scaling by f_l, is done as follows. If $f_l a_l$ is less than half the smallest division in the representation (1/16) it is approximated as 0 to avoid rounding up and increasing the output from irrelevant feature detectors. Otherwise, the approximation is

$$\overline{a_l} = \langle f_l\, a_l \rangle_{1/8,\,8} \quad (5.20)$$

selecting the closest power of 2 in the range. This results in 8 possible values, which can be encoded in 3 b; in fact, the encoding is compatible with the control signals required to implement the shifting. To summarize this very important result, *the weights a_l can be approximated with 3 bits*.

The power of 2 approximation of r_l is (applies only to the r_l multiplying the ϕ function):

$$\overline{r_l} = \langle r_l \rangle_{1,\ 16\ ;\ 2/3}. \tag{5.21}$$

This selects the closest power of 2 in the range, with a bias of 2/3 to rounding up i.e. if r_l is in the upper two-thirds of the range it is rounded up, in the lower one-third of the range it is rounded down. The bias is used to preferably increase rather than decrease S-cell output, once inhibition is properly handled as above.

5.3.4.2 DNC Model Functions

With the preprocessing approximations applied to the NC model S-cell output Equation (5.1), the DNC model S-cell output is

$$\overline{u_{Sl}(k,\ \underline{n})} = \overline{r_l}\ \phi \left[\frac{E - IC}{1 + IC} \right], \phi[x] = max(x, 0), \tag{5.22}$$

where the excitation sum is

$$E = \sum_{\kappa=1}^{K_{Cl-1}} \sum_{\underline{\nu} \in A_l} \overline{a_l(\kappa,\ \underline{\nu},\ k)}\ \overline{u_{Cl-1}(\kappa,\ \underline{n}+\underline{\nu})}, \tag{5.23}$$

and the inhibitory term is

$$IC = \overline{r_l b_l(k)/(1 + r_l)}\ \overline{v_{Cl}(\underline{n})}. \tag{5.24}$$

The inhibit (IC) portion is implemented as shown on the left of Figure 5.3. If $r_l b_l/(1 + r_l)$ is represented by a sum or difference of two powers of 2, the multiplication is a sum or difference of two shifts, implemented by the sign input to the accumulator to add or subtract. The DNC model approximates the $1/(1 + IC)$ inhibit function in Equation (5.22) by a function of the form $1/IC2$, where IC2 is a power of 2, to enable implementation by another shift operation. With this approximation, the S-cell output function is

$$\overline{u_{Sl}(k,\ \underline{n})} = \overline{r_l} \left[\frac{E - IC}{IC2} \right], \quad E > IC \tag{5.25}$$

$$= 0 \quad otherwise.$$

The IC2 approximation is defined as shown in Table 5.4 in decimal format; this can be implemented by simple combinatorial logic.

Table 5.4 Power of 2 Approximation for 1/(1+I) Inhibit Function

IC , IS Interval	IC2 , IS2 Approximation	Interpretation
< 0.5	1	shift 0
[0.5 , 2.0)	2	shift 1 R
[2.0 , 4.5)	4	shift 2 R
[4.5 , 10.0)	8	shift 3 R
[10.0 , 21.0)	16	shift 4 R
≥ 21.0	32	shift 5 R

The rest of the model is implemented as shown on the right of Figure 5.3. The same accumulator is used for the excitation sum of Equation (5.23) and the E - IC subtraction in Equation (5.25). To implement the ϕ function, the hardware must also detect $E - IC < 0$, and generate a force 0 signal to the following shifter.

5.3.5 Digi-Neocognitron: Inhibitory Vs-cells

Note that the presence of Vs-cells in the neural network model is an architectural option. Typically Vs-cells are not used on the lower levels, but are used on the final level to give a form of lateral inhibition.

5.3.5.1 *Preprocessing*

Equation 5.8 describes the inhibitory Vs-cell output function of the NC model. The fixed weights d_l are a two-dimensional Gaussian with variance σ_d and normalization $d_l(0) = 1$, and the same approximation (Table 5.1) as for c_l is used. The number K_{Sl} of previous level S-cell planes, assumed to be less than 128, is approximated as a power of 2 to replace the division in Equation (5.8) by a shift operation :

$$\overline{K_{Sl}} = \langle K_{Sl} \rangle_{1,\ 64\ ;\ le} . \tag{5.26}$$

5.3.5.2 DNC Model Functions

With the preprocessing approximations, the DNC model Vs-cell output is

$$\overline{v_{Sl}(\underline{n})} = \frac{1}{\overline{K_{Sl}}} \sum_{\kappa = 1}^{K_{Sl}} \sum_{\underline{\nu} \in D_l} \overline{d_l(\underline{\nu})} \, \overline{u_{Sl}(\kappa, \, \underline{n} + \underline{\nu})} \qquad (5.27)$$

which is implemented as shown in Figure 5.4. The final block approximates the $1/(1 + IS)$ inhibit function which occurs in the C-cell calculation (next section), by a function of the form $1/IS2$, with IS2 a power of 2 (Table 5.4).

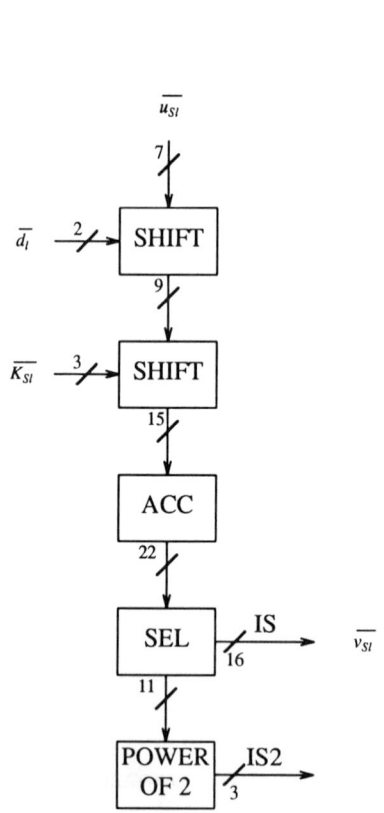

Figure 5.4 Digi-neocognitron Vs-cell block diagram, with bit widths indicated

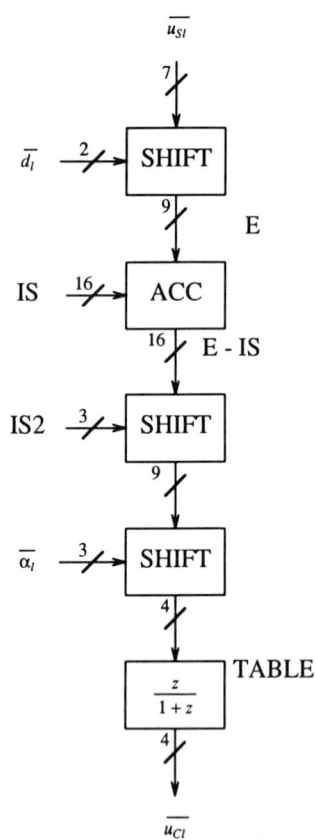

Figure 5.5 Digi-neocognitron C-cell block diagram, with bit widths indicated

5.3.6 Digi-Neocognitron: C-cells

5.3.6.1 Preprocessing

Equation 5.6 describes the C-cell output function of the NC model. The fixed weights d_l are approximated as before (Table 5.1). The saturation parameter α_l in Equation (5.7) is approximated as:

$$\overline{\alpha_l} = \langle \alpha_l \rangle_{1/32,\ 1\ ;\ le} . \tag{5.28}$$

The *le* rounding is used to increase rather than decrease C cell outputs.

5.3.6.2 DNC Model Functions

With the preprocessing approximations applied to the NC model C-cell output Equation (5.6), the DNC model C-cell output is

$$\overline{u_{Cl}(k,\ \underline{n})} = \psi \left[\frac{E - IS}{IS2} \right],\ E > IS \tag{5.29}$$

$$= 0 \quad otherwise,$$

where the saturation function ψ is defined in Equation (5.7), the excitation sum and inhibitory term are

$$E = \sum_{\underline{\nu}\ \epsilon\ D_l} \overline{d_l(\underline{\nu})}\ \overline{u_{Sl}(k,\ \underline{n}+\underline{\nu})},\quad IS = \overline{v_{Sl}(\underline{n})}, \tag{5.30}$$

and the $1/(1 + IS)$ inhibit function is approximated by $1/IS2$. This is implemented as shown in Figure 5.5. The final step in the DNC model involves approximation of the ψ function in Equation (5.7) in order to evaluate Equation (5.29):

$$\overline{\psi[x]} = \frac{z}{1+z},\ z = x/\overline{\alpha_l} . \tag{5.31}$$

This is implemented by a shift to transform approximately to z, followed by a table lookup implementation of the $\frac{z}{1+z}$ function, as defined in Table 5.5. The critical decision region for the saturation function Equation (5.31) is from $z = 1/3$, where the output is 0.25, to $z = 3$, where the output is 0.75; this allows defining the lookup table primarily to cover this range well, which only requires a 4-b input plus an overflow bit.

Table 5.5 Z/(1+Z) Table for C-Cell Output

Input (decimal)	Input (binary)	Output (binary)	Output (decimal)
0.0	00.00	0.0000	0.0
0.25	00.01	0.0011	0.1875
0.5	00.10	0.0101	0.3125
0.75	00.11	0.0111	0.4375
1.0	01.00	0.1000	0.5
1.25	01.01	0.1001	0.5625
1.5	01.10	0.1010	0.625
1.75	01.11	0.1010	0.625
2.0	10.00	0.1011	0.6875
2.25	10.01	0.1011	0.6875
2.5	10.10	0.1011	0.6875
2.75	10.11	0.1100	0.75
3.0	11.00	0.1100	0.75
3.25	11.01	0.1100	0.75
3.5	11.10	0.1100	0.75
3.75	11.11	0.1101	0.8125
>	>	0.1111	0.9375

5.3.7 Methodology for Conversion from NC Model to DNC Model

The procedure for conversion of an NC model to a DNC model utilizes the inherent NC fault tolerance characteristics [28] and unsupervised learning capability to compensate for inaccuracies introduced by the implementation approximations, and consists of the following steps.

1. Do unconstrained simulation (including learning) with the NC model, in order to get something that works for the particular problem. If there is not already an existing NC model, also incorporate the following step.

2. Change the NC model to use approximate fixed weights, as in Table 5.1. If feasible, also use appropriate powers of 2 for the selectivity r_l and saturation α_l parameters. This recognizes that some of the DNC model preprocessing approximations can be treated as just another set of NC model

An All-Digital VLSI ANN 177

parameters. Repeat simulation (including learning) with the NC model, and also tailor the number of training iterations to develop weights that are appropriate for the DNC representations, as in Equations (5.18), (5.19) and (5.20).

3. Preprocess level 1 to create level 1 DNC model approximations.

4. Run the DNC model simulator on level 1 for all input patterns in the training set, and compare to level 1 NC model outputs.

5. Repeat (if necessary) steps 2,3 and 4 in order to obtain good weight approximations, trying different scaling factors, or changing the amount of training.

6. Use the level 1 outputs from the DNC model to *retrain* level 2 of step 2 in the NC model. This is a very important step. The level 2 feature detectors *must* be developed (learned) with inputs from the lower levels that are representative of what will be seen in the *DNC* model. Otherwise, there will not be enough similarity between lower level outputs and higher level feature detectors to have a robust conversion of the NC model pattern recognition capabilities to the DNC model.

7. Preprocess level 2 to create level 2 DNC model approximations.

8. Repeat this process for all levels in the architecture e.g. if there are 3 or more levels, next run the DNC simulator on level 2, and use these outputs to retrain level 3 in the NC model.

9. Use the DNC model to check final level outputs for all input patterns using all levels in the neural network.

5.4 CHARACTER RECOGNITION EXAMPLE

This section reports simulation results on a character recognition example which proves the feasibility of the DNC model. It involves recognizing 7 by 12 pixel characters A through E left justified on a 12 by 12 pixel array [17]. The NC model is trained on *centered* versions of the input patterns (left side of Figure 5.6), both for generality and to test translation invariance in recognizing the left-justified patterns. The conversion methodology is followed to derive a DNC model, and both ANNs are tested on translations of the original patterns with varying levels of added random noise.

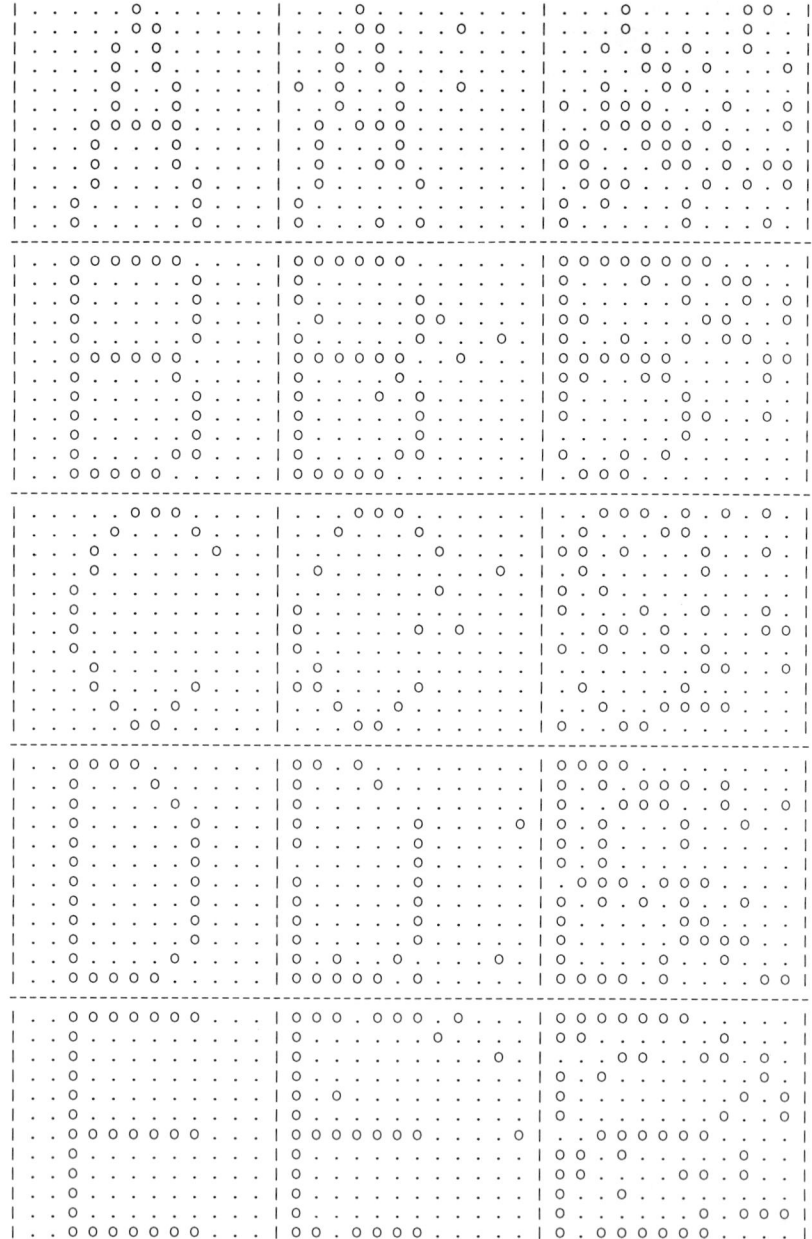

Figure 5.6 Character recognition example input patterns of 12 x 12 pixels: left, centered A through E used for training; middle, typical edge-translated A through E with 5% noise; right, typical edge-translated A through E with 30% noise

5.4.1 NC and DNC Model Solutions

The architecture and parameters of an NC model that provides a solution to this character recognition problem are described on the input file to the NC model simulator, as shown in the Appendix. Only 2 levels of S-cell and C-cell planes are required, as shown in Figure 5.1. The input plane is 12 by 12, and the 20 level 1 S-cell planes are oversized to 14 by 14, in order not to lose feature detector outputs when patterns are translated to an edge of the input plane. The number of neurons in the level 1 C-cell planes is thinned out by a 2:1 ratio (with oversizing) to 8 by 8, skipping every other C-cell since they carry nearly the same information due to *smearing* of the S-cell outputs through the 5 by 5 receptive field averaging. The 5 level 2 S-cell planes are thinned out, by dropping periphery locations which carry less information, to 6 by 6, and level 2 C-cells consist of one output neuron per pattern class. Vs-cell inhibition is used at level 2; this serves to sharpen distinctions between outputs but does not necessarily cause only one output neuron to be activated. In order to have good noise tolerance, it is necessary that more than one output neuron be activated for the noise-less version of some patterns, with output level in proportion to the strength of the recognition of the pattern e.g. consider the pattern E which is *part of* the pattern B in Figure 5.6.

This neural network architecture contains 5,762 neurons and 181,509 possible interconnections, with 2,800 unique weights possible (many may be zero). Training details are not presented here; there were only 7 iterations at level 1 and 14 iterations at level 2. When the NC model learning process is complete, only 910 (33%) of the possible 2,810 unique weights are non-zero. The NC model simulator also determines how many of the neurons and interconnections are active (non-zero) when a particular input pattern is processed; an interconnect is termed active if the weight is non-zero and the input neuron's activity is non-zero. For the five centered patterns A through E, the number of active neurons ranges from 321 (5.6%) to 539 (9.4%). The number of active connections ranges from 10,054 (5.5%) to 19,616 (10.8%). This is the basis for concluding that the Neocognitron is *temporally sparse* [16] as mentioned earlier.

The DNC model is derived from the NC model following the methodology outlined previously. The preprocessing parameter file for the DNC model, shown in the Appendix, uses a level 1 weight scaling factor of 0.991 to fine-tune the approximation, and represents the inhibitory weight factor with a single-term power-of-2 approximation. Level 1 of the DNC model is then run for all of the training patterns, and these outputs are used to repeat the level 2 NC model

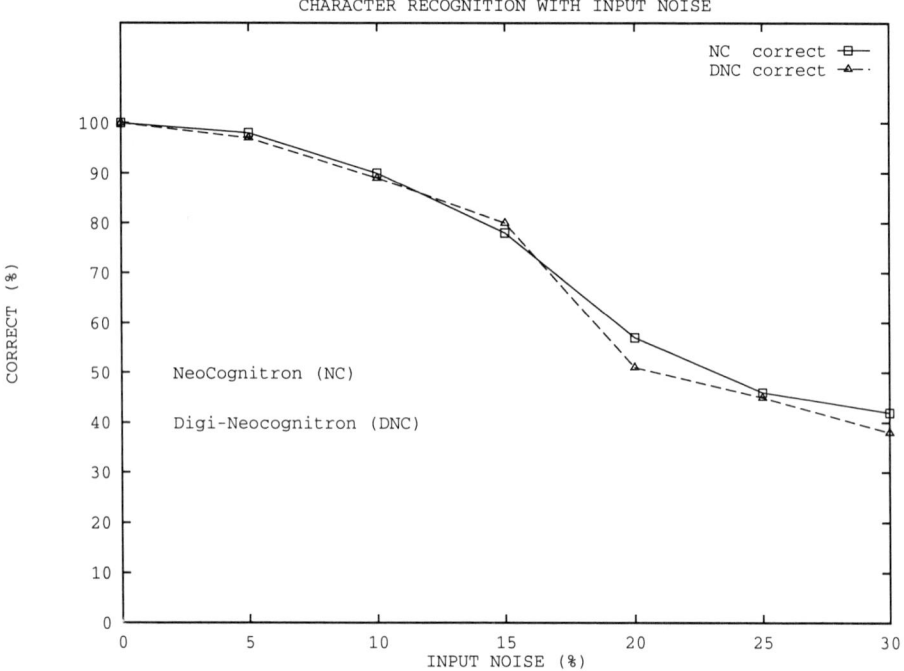

Figure 5.7 Neocognitron and Digi-Neocognitron character recognition: rejects as a function of input noise

learning. Level 2 is then preprocessed using a scaling factor of 0.949 and a two-term power-of-2 approximation for the inhibitory weight factors. The DNC model is then run on all patterns as a test. A final adjustment was done manually to slightly reduce the inhibitory weight factors for the B and C pattern S-cells since their level 2 C-cell outputs were low.

5.4.2 Results

Both ANNs are tested on translations of the original patterns with varying levels of random noise added; this quantifies recognition performance, translation invariance and noise tolerance and allows for a meaningful comparison of the NC and DNC models. As in [17], the original patterns are translated to the left edge, a fixed percentage of random pixel noise is added, and the noisy image is presented to the ANN for recognition. This test procedure is repeated 100 times at each noise level, in 5% increments ranging from 0% to 30%. Note that this random noise percentage is defined as a percent of the total number of pixels in the input plane, not in the pattern. At 5% noise, or a Hamming distance of 7 (i.e. 7 pixels are reversed), the patterns are still recognizable, as shown by the examples in the middle part of Figure 5.6. At 30% noise, or a Hamming distance of 43 (greater than the total number of pixels in any pattern), the patterns are not very recognizable, as shown by the examples in the right part of Figure 5.6.

The correct recognition rate as a function of noise added is shown in Figure 5.7. Both ANNs have excellent recognition rates in the presence of large amounts of noise, and the DNC model has no deterioration from NC performance.

The NC model recognition rate falls to 78% at 15% noise and is 42% when there is 30% noise; the DNC model recognition rate is 80% at 15% noise and 38% at 30% noise. These ANNs are also coping with translation and edge effects at the same time.

Both ANNs can also classify input patterns as unknowns or rejects, indicated by all C-cell outputs at zero or below a threshold, or by more than one near-equal maximum C-cell output. This significantly reduces the undetectable error rate (wrong classifications) and is very important for commercial applications [18]. The reject rate as a function of noise added is shown in Figure 5.8, and incorrect recognition rate as a function of noise added is shown in Figure 5.9. The NC model reject rate gradually rises, up to 11% at 15% noise and 38% at 30% noise. As a result, the incorrect recognition rate is less than 22%

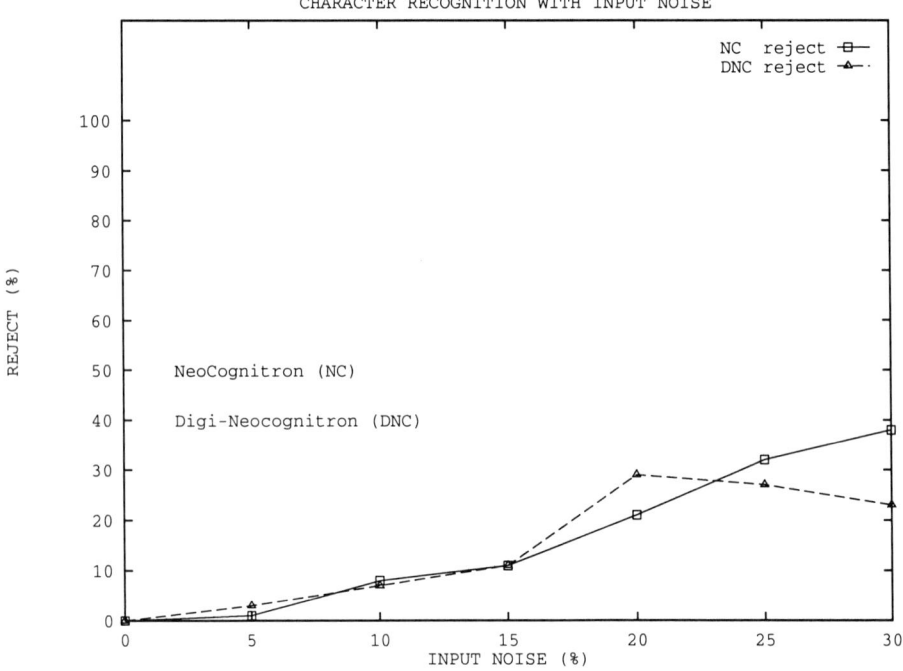

Figure 5.8 Neocognitron and Digi-Neocognitron character recognition: correct identifications as a function of input noise

(occurs at 20% and 25% noise) and is only 20% at 30% noise. The DNC model reject rate is 11% at 15% noise and 23% at 30% noise; the incorrect recognition rate is less than 39% (occurs at 30% noise). The NC and DNC model incorrect recognition rates are lower than Brown [17] at all input noise percentages. The ANNS were also tested on the shifted and original patterns, which were correctly recognized.

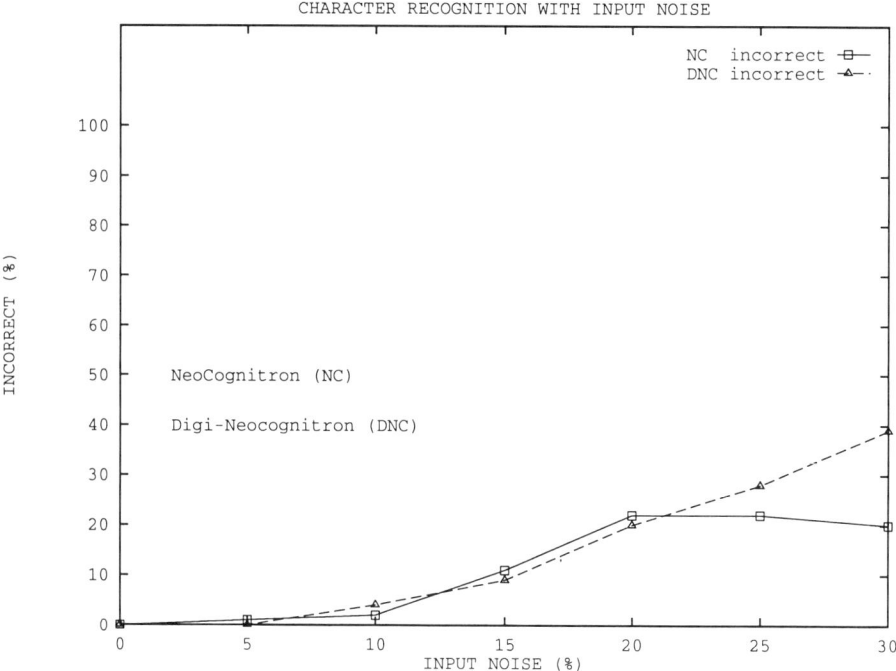

Figure 5.9 Neocognitron and Digi-Neocognitron character recognition: incorrect identifications as a function of input noise

5.5 ADVANTAGES FOR VLSI IMPLEMENTATION

Table 5.6 illustrates the advantages of the DNC model as compared to the NC model for VLSI implementation of the multiplication in the weighted sum calculations. An 8-bit parallel (array) multiplier is assumed for implementation of the NC model; this is formed from 48 full adders, 8 half adders and 64 AND gates, and has worst case delay of 17 times the worst case adder delay [29]. The DNC model implementation is as presented in Section 5.3.

The ACTEL ACT I family [30] is used as the Field-Programmable Gate Array (FPGA) technology example. The multiplier for the NC model requires 176 logic modules and has a worst case delay of 253 ns. The DNC model power of 2 shift circuit to implement $c_l u_C$ ($a_l u_C$) is designed using ACTEL MX4 4:1 multiplexer macros, and requires 4 (23) logic modules and has a worst case delay of 9.9 (19.2) ns.

Table 5.6 VLSI Advantage of Digi-Neocognitron Model

Model	NC	DNC	DNC
Calculation	$c_l u_C$ or $a_l u_C$	$c_l u_C$	$a_l u_C$
Implementation	multiplier	shifter	shifter
FPGA: Area (# modules)	176	4	23
Delay (ns)	253	10	19
CMOS gate array: Area (# cells)	810	31	133
Delay (ns)	61	3	5

The gate array technology example is Texas Instruments TGC100 1-micron CMOS gate arrays [31]. The multiplier for the NC model requires 810 gate array cells (adding 35% for routing) and has worst case delay of 60.8 ns. The DNC model power of 2 shift circuit to implement $c_l u_C$ ($a_l u_C$) is designed using gate array macros such as 2:1 and 4:1 muxes, and requires (31) 133 cells, assuming only 10% for routing since this is a much simpler circuit than the multiplier, and has worst case delay of 2.6 (4.6) ns.

The area-delay product is *improved by 2 to 3 orders of magnitude,* and this does not even consider the elimination of division and other complex functions. A custom CMOS implementation could achieve even better ratios by using a transmission gate implementation (barrel shifter) of the shift circuit. In addition, the restriction of the bit widths in the DNC model, such as 3-bit or less representation of weights, and representation of neuron outputs with 4 or 7 bits, substantially reduces the I/O and memory requirements.

5.6 CONCLUSIONS

This chapter has presented a new artificial neural network model, the Digi-Neocognitron (DNC), which adapts the Neocognitron (NC) neural network model to an efficient all-digital implementation for VLSI. It was shown how the DNC model is derived from the NC model by a combination of preprocessing approximations and definition of new model functions. For example, multiplication and division operations are eliminated by conversion of factors to powers of 2, so that only shift operations are required. A methodology for conversion of NC models to DNC models was presented. The feasibility of this approach was demonstrated on a character recognition example. The DNC model had the same pattern recognition performance as the NC model, when tested on translated and noisy input patterns. The DNC model has substantial advantages over the NC model for VLSI implementation. The area-delay product is improved by 2 to 3 orders of magnitude, and I/O and memory requirements are reduced by representation of weights with 3 bits or less and neuron outputs with 4 bits or 7 bits.

APPENDIX
Architectural Description File for NC Model

lmax 2
layer U0 12 12

layer US1 14 14 20 nothin
selectivity 5.0
reinforcement 1.0
connection area 3 3 approx
sigmac 3.0

layer UC1 8 8 20 thin normal
saturation 0.25
connection area 5 5 approx
sigmad 2.5

layer US2 6 6 5 thinper
selectivity 1.0
reinforcement 24.0
connection area 5 5 approx
sigmac 5.0

layer UC2 1 1 5 thinper inhibit
saturation 0.4
connection area 6 6 approx
sigmad 8.0

Preprocessing Parameter File for DNC Model

wscale 1 0.991
wscale 2 0.949

brpow2 1 1
brpow2 2 2

REFERENCES

[1] Rumelhart, D.E. and McClelland, J.L., Parallel Distributed Processing

Volume 1: Foundations, Cambridge: The MIT Press, 1986.

[2] Lippmann, R.P., "An introduction to computing with neural nets," IEEE ASSP Magazine, vol. 4, pp. 4-22, Apr. 1987.

[3] Hecht-Nielsen, R., "Neurocomputing: picking the human brain," IEEE Spectrum, vol. 25, pp. 36-41, Mar. 1988.

[4] IEEE Trans. Neural Networks, 1990.

[5] Atlas, L.E. and Suzuki, Y., "Digital systems for artificial neural networks," IEEE Circuits and Devices Magazine, pp. 20-24, Nov. 1989.

[6] Treleaven, P., et al., "VLSI architectures for neural networks," IEEE MICRO Magazine, pp. 8-27, Dec. 1989.

[7] Proc. IEEE IJCNN, San Diego, 1990.

[8] Howard, R.E., et al., "Optical character recognition: a technology driver for neural networks," Proc. IEEE ISCAS, 1990, pp. 2433-2436.

[9] Jackel, L.D., et al, "Hardware requirements for neural-net optical character recognition," Proc. IEEE IJCNN, San Diego, 1990, vol. II, pp. 855-861.

[10] Hammerstrom, D., "A VLSI architecture for high-performance, low-cost, on-chip learning," Proc. IEEE IJCNN, San Diego, 1990, vol. II, pp. 537-544.

[11] Tomlinson, M.S., et al., "A digital neural network architecture for VLSI," Proc. IEEE IJCNN, San Diego, 1990, vol. II, pp. 545-550.

[12] Gamal, A El., et al, "An architecture for electrically configurable gate arrays," IEEE Journal of Solid-State Circuits, vol. 24, no. 2, pp. 394-398, Apr. 1989.

[13] Fukushima, K., "Neocognitron: a self-organizing neural network model for a mechanism of pattern recognition unaffected by shift in position," Biol. Cybernetics, vol. 36, pp. 193-202, 1980.

[14] Fukushima, K. and Miyake, S., "Neocognitron: a new algorithm for pattern recognition tolerant of deformations and shifts in position," Pattern Recognition, vol. 15, no. 6, pp. 455-469, 1982.

[15] Fukushima, K., "Analysis of the process of visual pattern recognition by the neocognitron," Neural Networks, vol. 2, pp. 413-420, 1989.

[16] Hammerstrom, D. and Means, E. "System design for a second generation neurocomputer," Proc. IEEE IJCNN, Jan, 1990, vol. II, pp. 80-83.

[17] Brown, H.K., et al., "Orthogonal extraction training algorithm," Proc. IEEE IJCNN, Jan, 1990, vol. I, pp. 537-540.

[18] Wilson, C.L., et al, "Self-organizing neural network character recognition on a massively parallel computer," Proc. IEEE IJCNN, San Diego, 1990, vol. II, pp. 325-329.

[19] Fukushima, K. Miyake, S. and Ito, T., "Neocognitron: a neural network model for a mechanism of visual pattern recognition," IEEE Trans. Systems, Man, and Cybernetics, vol. SMC-13, no. 5, pp. 826-834, Sep. 1983.

[20] Lee, Y., et al, "Hangul recognition using neocognitron," Proc. IEEE IJCNN, Jan, 1990, vol. I, pp. 416-419.

[21] Wang, S. and Pan, C. "A neural network approach for Chinese character recognition," Proc. IEEE IJCNN, San Diego, 1990, vol. II, pp. 917-923.

[22] Barnard, E. and Casasent, D. "Shift invariance and the neocognitron," Neural Networks, vol. 3, pp. 403-410, 1990.

[23] Johnson, K., et al, "Feature extraction in the neocognitron," Proc. IEEE IJCNN, 1988, vol. II, pp. 117-126.

[24] White, B.A. and Elmasry, M.I., "Digi-Neocognitron : a Neocognitron neural network model for VLSI", Report No. UW/VLSI 91-01, VLSI Group, University of Waterloo, Waterloo, Ontario, Jan, 1991.

[25] White, B.A. and Elmasry, M.I. "VLSI architecture of the Digi-Neocognitron neural network," to be published.

[26] Lim, Y.C., et al, "VLSI circuits for decomposing binary integers into signed power-of-two terms," Proc. IEEE ISCAS, 1990, pp. 2304-2307.

[27] Marchesi, M., et al, "Multi-layer perceptrons with discrete weights," Proc. IEEE IJCNN, San Diego, 1990, vol. II, pp. 623-630.

[28] Xu, Q., et al, "A fault tolerance analysis of a neocognitron model," Proc. IEEE IJCNN, Jan, 1990, vol. II, pp. 559-562.

[29] Weste, N.H.E. and Eshraghian, K., Principles of CMOS VLSI Design : A Systems Perspective, p. 344. Reading : Addison-Wesley Publishing Company, 1985.

[30] ACTEL Corporation, "ACT 1 Family Gate Arrays Design Reference Manual", 1989.

[31] Texas Instruments, "TGC100 Series 1-micron CMOS Gate Arrays", May, 1990.

6

A NEURAL PREDICTIVE HIDDEN MARKOV MODEL ARCHITECTURE FOR SPEECH AND SPEAKER RECOGNITION

Khaled Hassanein, Li Deng and Mohamed I. Elmasry

6.1 INTRODUCTION

Speech is undoubtly the most natural and efficient form of human communication. For several decades researchers in the field of automatic speech recognition (ASR) have been driven with the goal of establishing a means through which people can talk to computers in much the same way as they carry on conversations with fellow humans.

The potential applications of such a technology are endless. Some of the main applications include computerized dictation systems, voice operated telephone dialing, voice control of machines and voice reservation systems for travel or purchasing.

Although some breakthroughs have been achieved in the area of automatic speech recognition, this success seems really limited if we compare the abilities of current ASR machines to those of a five year old child.

The main reasons for the difficulties faced by machines in this regard stem from the following facts:

(i) The speech signal contains a high level of variability. This variability is present on the same speaker level, depending on his speaking mode (e.g. relaxed, stressed, singing, etc.). It is also present between different speakers depending on several factors (e.g. age and sex). It is also subject to the surrounding environment (e.g. noise level).

(ii) The basic units of speech (phonemes) are pronounced differently in different context. This is referred to as the *coarticulation* problem. Thus, the same phoneme could sound differently based on both its close context, (i.e., phonemes directly preceding and proceeding it), and its long term context, (i.e., its place in the sentence).

(iii) The speech signal contains various sorts of information, (i.e., sounds, syntax, semantics, sex, and personality). This requires the ASR system to perform analysis in all of the above directions in order to arrive at correct decisions.

(iv) Very high computation rates are required to handle all of the above cues simultaneously and perform the type of analysis required for real-time speech recognition.

(v) The lack of a comprehensive theory of speech recognition. Presently, there are no rules for formalizing the information at the various levels of speech decoding.

(vi) The problem of continuous speech recognition is further complicated by the fact that there are no silence gaps between words spoken naturally.

For all of the above reasons, the best systems available today for automatic speech recognition perform well only under various constraints. These include speaker dependency, isolated words, low noise environment and small vocabulary size.

6.2 AUTOMATIC SPEECH RECOGNITION METHODOLOGIES

Having outlined the main problems associated with the task of automatic speech recognition, in this section we review some of the methods that have achieved marked progress in solving some of these problems and have thus resulted in a significant improvement in the performance of automatic speech recognition systems that exist today.

Three basic approaches are covered in this section. They represent the most effective methods used for speech recognition at the present time. These approaches are:

(i) The statistical approach (namely, the Hidden Markov Model (HMM)) [1, 2, 3, 4]

(ii) the Artificial Neural Network (ANN) approach [5, 6, 7, 8, 9, 10, 11].

(iii) The Hybrid approach combining both the above methods [8, 9, 11, 12, 13, 14, 15, 16, 17, 18, 19, 20, 21, 22, 23, 24].

The following subsections explain these methods in detail and provide a discussion of their advantages and disadvantages.

6.2.1 Hidden Markov Models

As explained in Section 1 above, speech differs from most signals dealt with by conventional pattern classifiers in that the information is conveyed by the temporal order of the speech sounds. A stochastic process provides a way of dealing with both the temporal structure and the variability within speech patterns representing the same perceived sounds. One of the most popular stochastic approaches to speech recognition today is statistical learning, based on HMMs. In this type of model each word of the lexicon is modeled by a Markov chain [3]. This method is explained in detail below.

6.2.1.1 Basic Structure

An HMM is a group of states linked together by transitions associated with a transition probability. The states are not directly observable in the sense that the observations for each state are specified by a Probability Density Function (PDF) rather than by deterministic mechanisms. A typical HMM is shown in Figure 6.1.

An HMM could be defined as follows:

(S): A collection of states that includes an initial state S_i and a final state S_f.

(a_{ij}): A set of transition probabilities of taking a transition from state i to state j. If we are dealing with an ergodic model, i and j could assume any values. However, when dealing with speech, frequently a left to right model is used to model the speech sequence and therefore, the condition of ($j \geq i$) is imposed.

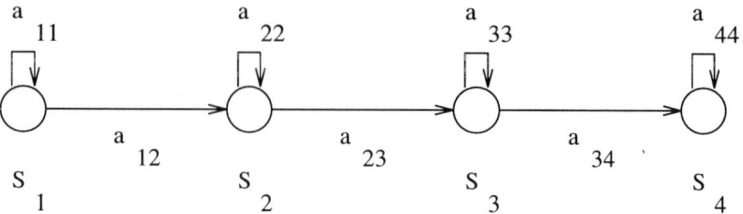

Figure 6.1 A 4 state left to right HMM

$(b_i(F))$: A set of observational probabilities or the probability of observing data F when visiting state i.

The Markov model is said to be hidden, since one cannot directly observe which state the Markov model is in. Only the data generated by the states can be directly observed. Since both a and b are probabilistic in nature, they must satisfy the following conditions:

$$a_{ij} \geq 0, \ b_i(F) \geq 0, \ for \ all \ i, \ j. \tag{6.1}$$

$$\sum_j a_{ij} = 1, \ for \ all \ i \tag{6.2}$$

$$\sum_F b_i(F) = 1, \ for \ all \ i \tag{6.3}$$

In a first order HMM, two basic assumptions are held:

(i) The Markov assumption, which simply states that the probability of the Markov model being in state i at time t only depends on the state it visited at time $t-1$.

(ii) The output independence assumption, which means that the data emitted by state i at time t depends only on the transition taken at that time (from S_{t-1} to S_t).

These assumptions impose limitations on the memory of the model, however, they reduce the number of parameters and consequently make learning and decoding very efficient.

There are three basic problems to be solved in hidden Markov modeling. These could be stated as follows [1]:

(i) The *Evaluation Problem:* What is the probability of a given model generating a given sequence of events? This can be solved by the *Forward* algorithm [3], which gives the maximum likelihood estimation that the given sequence was generated by the model.

(ii) The *Decoding Problem:* Given a model and a sequence of events, what is the most probable state sequence in the model that would have resulted in this sequence of observations? This problem is efficiently solved by using the *Viterbi* algorithm [25].

(iii) The *Training (Learning) Problem:* How to optimize the parameters of the model so as to maximize its probability of generating a certain sequence of events. This can be done through the use of the *Forward-Backward (Baum-Welsh)* algorithmin the case of the Maximum Likelihood (ML) training scheme [26].

The details of the above algorithms are described in [1, 3] and will not be described here.

There are two basic types of HMMs:

(i) *Discrete HMMs:* In this type of HMMs each acoustic unit to be recognized (e.g. words) is represented by a Markov model as described above. The discrete HMM is characterized by the use of vector quantization. Here, similar training events are grouped into clusters and each cluster is represented by only one observation probability. Therefore, the output distributions are non-parametric and are determined by the output probabilities stored in code books.

(ii) *Continuous HMMs:* In this type, the output distributions are parametrized. The discrete output probabilities are replaced by a model of the continuous spectrum of the data space. Thus, it allows for more accuracy. A popular model used for continuous HMMs is the Multivariate Gaussian Density model, which describes the PDF of a mean vector and a covariance matrix. Also, a mixture of Gaussian densities could be used to yet improve the accuracy [27, 28].

Several comparisons have been conducted between discrete and continuous HMMs. Only complex continuous models, requiring much more computation times, proved to be better than discrete HMMs [29, 30].

6.2.1.2 Training Criteria

Several criteria have been utilized to train HMMs for speech recognition applications. The most commonly used of these criteria are summarized below:

Maximum Likelihood (ML) Training:

This is the most widely used algorithm for the estimation of HMM parameters to fit a given sample of labeled training data. ML training is based mainly on the iterative Estimation-Maximization (EM) algorithm principle [31] in which a new model is computed that is guaranteed to improve the probability of the observation sequence upon each iteration. The EM algorithm, when applied to an HMM, is known as the Baum-Welsh Algorithm [26]. ML guarantees optimal training only if the model used is the correct one, which is rarely the case.

Corrective Training:

This method could be summarized as follows [32]:

(i) Use part of the training data and ML training to obtain an initial estimate of the model parameters θ.

(ii) Carry out speech recognition on the whole training data set using the model.

(iii) If any word W is misrecognized as w, adjust the parameters of the model θ so as to make W more probable and w less probable.

(iv) If θ has been changed in step (iii), return to step (ii).

Corrective training will try to find parameters that improve the model performance, rather than just trying to maximize the likelihood as is the case in ML training. Also, since corrective training starts with initial ML estimates and since any parameter vectors that increase the error rate are discarded, the final estimates, with corrective training, should be better than those obtained using ML. However, minimizing the error rate with respect to the training data does not guarantee the same for testing data unless the operational conditions of the test data are similar to those of the training set. Therefore, it is of utmost importance to carefully choose the training data set as a true representation of the application desired.

Maximum Mutual Information (MMI) Training:

This approach is similar to corrective training but is mathematically more rigorous. In MMI training, the objective is to optimize the model parameters θ so as to maximize the likelihood of generating the acoustic data, given the correct word sequence as in ML. However, at the same time the probability of generating any wrong word sequence is minimized. While training is performed independently for each model when using criteria (1 or 2) above, with MMI training each training sample must be fed to all models simultaneously and optimization is performed for each model based on information from all models. Compared to ML, MMI training is more robust when the assumed model is incorrect [29] and generally yields better results [33, 34].

6.2.1.3 Speech Modeling Units

HMMs can be used to model any unit of speech. Several schemes have been investigated. Some of these schemes are summarized below:

Word Units:

This is the most natural unit of speech, and has been extensively used. Although our ultimate goal is to recognize words, they pose a number of problems in HMM modeling of speech. The problem is that in order to get an accurate word model there should be a large number of training data for that particular word. In large vocabulary recognition this presents a real problem because of the amount of training and storage required. Also the recognition process considers a word as a whole, and does not focus on the information distinguishing two acoustically similar words. On the other hand, word models have the ability to internally associate within-word coarticulatory effects and hence the acoustic variability at the subword level will not cause any additional recognition error.

Phoneme Units:

A solution to the computationally prohibitive task required for word based HMMs is to use subwords as the modeling units. Using phonemes as the basic modeling units offers two main advantages:

(i) It allows a high degree of data sharing between different words, thus alleviating the problems associated with having to deal with large training sets. This makes it very well suited for large vocabulary recognition tasks, since there is only a fixed number of phonemes regardless of the lexicon size.

(ii) Discrimination between words should improve, since now in similar words, only the relevant differences will be effectively represented. This is unlike the complete word-model, where accidental variations in the common parts of two similar words will likely result in recognition errors.

However, one major disadvantage of this technique is its inability to capture the acoustic variability associated with phonemic segments in different contexts. This variability can be averaged out, and with increased training, the system's performance will quickly saturate [35].

Microsegment Units:

The main idea here is to reduce the contextual effects through the use of many sub-phonemic small speech segments selected to be acoustically consistent according to human knowledge. Microsegments are at least as trainable as phonemes. Data sharing is achieved across different phonemes rather than just across different words as in phoneme units. This method, however, suffers from two major drawbacks [36]:

(i) Since there exists a wide range of phonemical variations, the choice of microsegments and construction of lexical items based on them is a difficult task for large vocabulary systems.

(ii) Certain microsegments have to be made context-dependent no matter how finely we cut speech into microsegments. This results in making the data sharing scheme very rigid and thus increases the training data required as in the word model.

Phonetic Feature Units:

To overcome the problems of the microsegment approach, the phonetic feature model was proposed [37]. Here, all possible intermediate articulatory configurations in transition from one target to the next are represented by selected combinations of phonetic feature values.

6.2.1.4 Drawbacks of The HMM Approach

Although HMMs have provided a good means for representing the complicated characteristics of the speech signal, they suffer from several shortcomings [13]. These are described below:

(i) Poor discrimination due to the ML training algorithm which maximizes the maximum likelihood estimate rather than the Maximum Apriori Probability (MAP). Although MMI training provides more discriminative power, the mathematics is more involved and constraining assumptions have to be made.

(ii) The assumptions made regarding the model structure and statistical distributions might often be inappropriate.

(iii) The assumption that the proper state sequence is governed by a first order hidden Markov chain is inaccurate most of the time.

(iv) For the standard HMM, correlations known to exist between successive acoustic vectors are lost. The inclusion of acoustic, phonetic or contextual information requires a more complex HMM for which the efficient Baum-Welsh algorithm often can not be applied directly. Also, a prohibitive storage capacity would be required.

6.2.2 Artificial Neural Networks for Speech Recognition

The complexity and nature of speech recognition has made it a good candidate for Artificial Neural Network (ANN) research. Recently, there has been a surge of research into the area of speech recognition using ANNs. Excellent reviews to this area have appeared in [8, 9, 11]. Here, we will attempt to give a quick overview of some of the most successful applications of Anns in the area of speech recognition.

The Multi-Layer Perceptron (MLPs) shown in Figure 6.2 has been successfully utilized in many speech recognition paradigms. The advantages of this approach include [13]:

(i) MLPs have strong discriminative abilities. It has been shown that MLPs, with two layers of weights (i.e. one hidden layer), can approximate any kind of nonlinear discriminant function [38].

(ii) Through nonlinear and non-parametric regression, MLPs are able to extract input data, relevant to classification without making any assumptions about the underlying statistical distributions.

(iii) With an appropriately sized training set, these models will exhibit generalization properties. These interpolative capabilities could then be used to perform statistical pattern recognition over an undersampled pattern space, without restrictive simplifying assumptions [20].

(iv) Their highly parallel architecture facilitates efficient real-time implementations.

(v) Finally, their flexible structure permits dealing with contextual inputs or recurrence quite easily.

The main ANN methods used for automatic speech recognition are reviewed below.

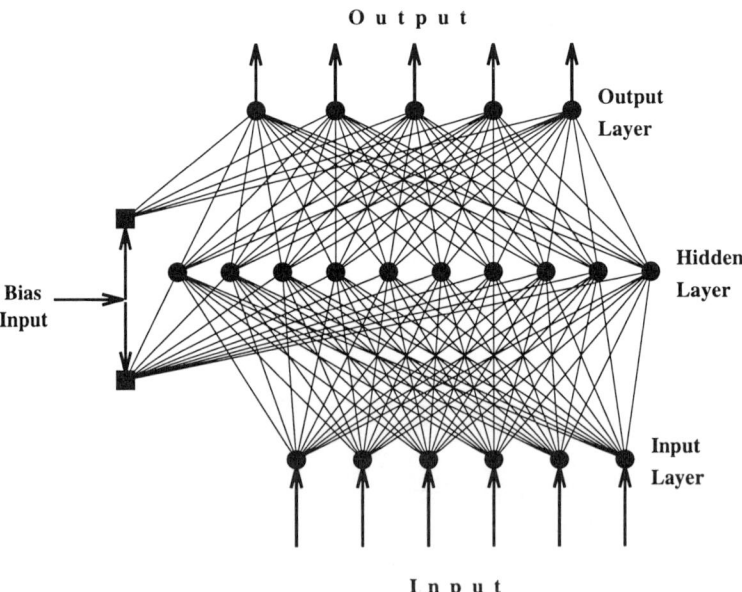

Figure 6.2 A Feed-forward MLP artificial neural network

6.2.2.1 Static Classification

Several network classifiers were used to classify static speech input patterns. These input patterns are usually based on a spectral analysis of pre-segmented words, phonemes or vowels. In this class of classifiers, the input at each word is accepted once as a whole static spectrographic pattern. The neural networks used here are static in nature (i.e. do not include internal delays or recurrent connections that could utilize the temporal nature of speech). This approach has the advantage of being able to capture the coarticulation effects in the different speech segments of the same word. However, it can not be used for real-time speech recognition, since it exhibits long delays to carry out the segmentation required. It also requires accurate pre-segmentation of both training and testing data for good performance. This could prove very costly, since it is usually done by hand [8].

Multilayer perceptrons, hierarchical networks (e.g. Kohonen's Learning Vector Quantizer (LVQ) and the feature map classifier) have been used extensively in this recognition scheme [6, 7, 39]. Results obtained from these studies indicate excellent performance on isolated-word speaker dependent tasks that compare well to the performance of HMM based recognizers.

6.2.2.2 Dynamic Classification

Dynamic neural networks classifiers have been developed to accommodate the dynamic nature of speech signals in the time domain. These networks mainly incorporate short time delays [40], or recurrent connections [12]. Unlike static networks, spectral inputs to dynamic networks are not given at once but rather fed sequentially frame-by-frame to the input units of these networks. Therefore, these dynamic classifiers could be used for real-time speech recognition. This is due to the fact that pre-segmentation is no longer needed.

Multilayer perceptron with finite time delays and networks with recurrent connections have been used extensively for the recognition of isolated words, digits and phonemes. These two methods are described briefly below.

(i) Recurrent Neural Nets: Recurrent nets [12] are a class of neural networks that employ feedback from the output of the hidden or output layers to the input layer. This could be viewed as introducing some sort of short term memory to the network. Hence, in making decisions the network is not only guided by the current input, but also utilizes the previous output. This scheme is particularly very attractive for speech recognition due to the coarticulation effects present between successive parts of the speech signal. An instance of recurrent neural networks is shown in Figure 6.3.

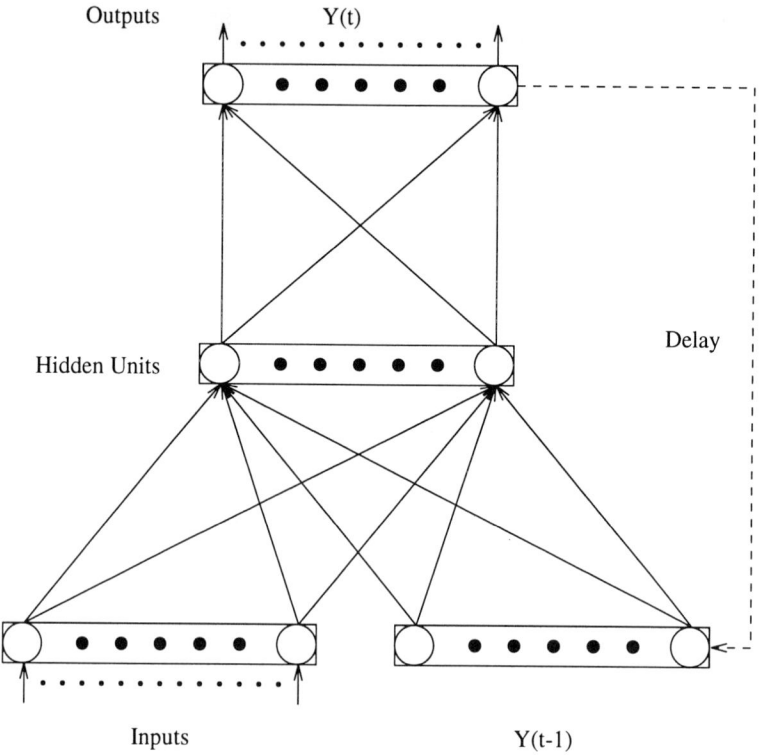

Figure 6.3 A Recurrent artificial neural network

(ii) Time Delay Neural Networks (TDNN): This method [10, 40] has found wide spread use in the area of isolated word recognition. It approximates a recurrent ANN over a finite time period by replacing the loops at each layer of the MLP by the explicit use of several preceding activation values. In this case the activations in a particular layer are computed from the current and the multiple delayed values of the preceding layer. Although this method has shown great promise for speech recognition, it clearly lacks some of the most important HMM features such as time warping. It also requires large sets of hand labeled training data and impressive training times.

6.2.3 The Hybrid Approach

As described above, the HMM approach to speech recognition has proven to be very successful in dealing with the dynamic nature of the speech signal. However, this approach suffers from a low discriminative power that limits its performance especially with highly confusable words.

On the other hand, ANNs, whether utilizing a static or a dynamic approach exhibit high discriminative abilities, but lack the proper structure for dealing with the dynamic nature of the speech signal. Therefore, it was quite natural for researchers to try integrating these two approaches to capitalize on their advantages in order to come up with more efficient and powerful ASR systems [8, 9, 11, 12, 13, 14, 15, 16, 17, 18, 19, 20, 21, 22, 23, 24].

In the remaining sections of this chapter we will present a hybrid ASR system based on a neural predictive HMM architecture [15, 16]. This class of predictive models will be analyzed in detail. Results on speech recognition experiments using this model are also presented. Various criteria are evaluated for training this class of models [41, 42]. Finally we show how the neural predictive HMM was successfully employed for implementing a fixed text speaker recognition system [43].

6.3 AN ANN ARCHITECTURE FOR PREDICTIVE HMMS

Speech frames generated by speech preprocessors in automatic speech recognizers typically possess strong correlations over time [44, 45]. The correlations stem, to a large degree, from the complex interactions and overlap patterns among various articulators involved in the dynamic process of speech production [46]. Standard hidden Markov models, based on the state-conditioned independent and identical distribution (IID) assumption [26], are known to be weak at capturing such correlations. The strength of data correlations in the HMM source decays exponentially with time due to the Markov property, while the dependence among speech events does not follow such a fast and regular attenuation.

The linear predictive HMM proposed in [2] and [47] was intended to overcome this weakness but shows no clear evidence of superiority over the standard HMM in speech recognition experiments [47]. This can be understood because the correlation (or the envelop of the correlation function) introduced by the state-dependent linear prediction mechanism decays also in an exponential manner with time lag [48]. This makes the capability of the linear predictive HMM, in dealing with speech-frame correlations, essentially the same as that exhibited by a standard HMM having just a larger number of states.

Nonlinear time series models [49, 50] are believed to be capable of representing the temporal correlation structure of speech frames in a more general and realistic manner. In order to represent the well known nonstationary nature of speech frames, the parameters of time series models can be made to vary with time. One elegant way of achieving this is to assume that the evolution of the time series model parameters follows a Markov chain.

In this section we describe an implementation of this idea where three-layered feed-forward neural networks are used as Markov-state-dependent nonlinear autoregressive type functions in a time series model. This gives rise to the Neural Predictive HMM (NP-HMM) which have been explored in detail in this study.

Layered neural networks are ideal tools for implementing mapping functions applicable to speech-frame prediction for two main reasons. First, it has been proved that a network of just one hidden layer is sufficient to approximate arbitrarily well any continuous function [51, 52]. Thus prediction of highly dynamic and complex speech frames can be potentially made as accurate as necessary. Second, the Back-propagation algorithm [53] is available for network parameter estimation.

Neural predictive models have been recently utilized by many researchers in the area of speech recognition [18, 19, 23]. To understand the properties of these predictive models, we carried out detailed analysis on the statistical correlation structures of various first-order predictive models. One principal conclusion drawn from the result of this analysis indicates that long-term temporal correlations, in the modeled data, cannot be achieved with linear prediction or with compressive nonlinear prediction using a single-predictive term. However, combinations of linear and compressive nonlinear terms are shown, analytically and by simulation, to be able to produce such signal correlations, which is a desirable property for a speech model. Speech recognition experiments conducted on a speaker-dependent discrete-utterance E-set recognition task, with various types of predictive HMMs, demonstrate close relationships between the recognition accuracy and the capabilities of these models in handling temporal correlations of speech data.

6.3.1 Correlation Structure in Speech Data and Coarticulation in Speech Dynamics

Speech patterns are known to be highly dynamic and complex in nature [46]. One principal source of this complexity is coarticulation. In articulatory terms, coarticulation results from the fact that several articulators do not always move instantaneously and simultaneously from one targeted articulatory configuration to another. In acoustic terms, coarticulation is related to context dependence, whereby acoustic realization of a sound is strongly affected by the sounds just uttered and those to be uttered next. This context dependence makes any IID source model, or a local IID source model as is the case with the standard HMM, a poor choice for fitting speech data and is a major source of errors in speech recognition [54]. Good speech models should provide correlation structures rich enough to accommodate the context dependence and other types of temporal dependence in speech data.

The correlation function for a random discrete-time source Y_t is defined as the normalized autocovariance function

$$\gamma_t(\tau) = Cov(Y_t, Y_{t+\tau}) = E[(Y_t - E(Y_t))(Y_{t+\tau} - E(Y_{t+\tau}))] \quad (6.4)$$

Here we assume Y_t is second-order stationary but can be nonstationary in means as in a recently developed model for speech [55]. The theoretical significance of the above covariance function is that it completely characterizes the second-order statistics of a random process. Intuitively, we can also argue easily for close relationships between the γ function and speech frame dependence as follows: If $\gamma(\tau)$ has large values at τ, then the acoustic data in a speech frame would have strong influences on those in another speech frame which is τ frames away. For instance, suppose $\gamma(\tau)$ has a large positive value; then if Y_t is greater than the mean value, $Y_{t+\tau}$ would tend to be forced to move above the mean value so as to keep $\gamma(\tau)$ positive. This is the kind of speech-frame dependence we would expect any reasonably good speech model to be able to handle.

6.3.2 Analysis of Correlation Structures for State-Conditioned Predictive Models

In this subsection, we conduct an analytical evaluation of the correlation functions for various types of predictive models (linear, compressively nonlinear, and their combination) in order to assess their faithfulness as a speech model for use in speech recognition. For the sake of simplicity, we assume first-order prediction and scalar observations. In neural network terminology, the analysis in this subsection concerns only one single neuron with no layered structure. We also omit the state label for the parameters of all the models for simplicity.

6.3.2.1 Linear Prediction

The state-conditioned linear predictive source model for speech data Y_t's is chosen to have the following form

$$Y_{t+1} = \phi Y_t + \epsilon_{t+1}, \quad t = 0, 1, ..., T. \tag{6.5}$$

where ϵ_t is an IID residual random variable with zero mean and variance σ^2 and the predictive function is a linear function of the data.

It is well known that when the predictive coefficient ϕ is less than one in absolute value, then the process represented by Equation (6.5) is stationary and its autocovariance function is given by

$$\gamma(\tau) = \frac{\sigma^2}{1 - \phi^2} \times \phi^\tau \tag{6.6}$$

That is, the autocovariance function declines exponentially fast as the time lag τ increases [48].

6.3.2.2 Prediction with a Compressive Nonlinear Function

The state-conditioned nonlinear predictive source model to be explored replaces the linear predictive term in Equation (6.5) with a compressive nonlinear function $f(\cdot)$ which is continuously differentiable

$$Y_{t+1} = f(Y_t) + \epsilon_{t+1}, \quad t = 0, 1, ..., T. \tag{6.7}$$

In actual implementation of this predictive model, $f(\cdot)$ is chosen as a specific compressive nonliner function, such as the tanh function given by

$$f(Y_t) = \frac{exp(Y_t) - exp(-Y_t)}{exp(Y_t) + exp(-Y_t)} \tag{6.8}$$

A Neural Predictive HMM for Speech Recognition

This is one of the standard nonlinear functions used in neural nodes in a wide class of artificial neural networks.

The method we use to derive the correlation function for Equation (6.7) resembles the perturbation analysis for the study of nonlinear differential Equations [56]. To proceed, we construct a family of models which are parameterized by α

$$Y_{t+1}(\alpha) = \alpha f(Y_t(\alpha)) + \epsilon_{t+1} \qquad (6.9)$$

and the model of Equation (6.7) is considered as one model in the family of Equation (6.9) whose statistical properties change continuously with the parameter α.

Once the model is parameterized, the autoregression on the data Y_t can be removed by performing a power-series expansion on the nonlinear function $f(\cdot)$

$$\begin{aligned}
Y_1(\alpha) &= \epsilon_1 + \alpha f(Y_0(\alpha)) \\
Y_2(\alpha) &= \epsilon_2 + \alpha f(Y_1(\alpha)) \\
&= \epsilon_2 + \alpha f[\alpha f(Y_0(\alpha)) + \epsilon_1] \\
&= \epsilon_2 + \alpha f(\epsilon_1) + \alpha^2 f(Y_0) f'(\epsilon_1) + \frac{1}{2}\alpha^3 f^2(Y_0) f''(\epsilon_1) + \cdots \\
Y_3(\alpha) &= \epsilon_3 + \alpha f(Y_2(\alpha)) \\
&= \epsilon_3 + \alpha f[\epsilon_2 + \alpha f(\epsilon_1) + \alpha^2 f(Y_0) f'(\epsilon_1) + \cdots] \\
&= \epsilon_3 + \alpha f(\epsilon_2) + \alpha^2 f(\epsilon_1) f'(\epsilon_2) + \alpha^3 f(Y_0) f'(\epsilon_1) f'(\epsilon_2) + \cdots \\
&\vdots
\end{aligned}$$

and in general,

$$Y_t(\alpha) = \epsilon_t + \alpha f(\epsilon_{t-1}) + \alpha^2 f(\epsilon_{t-2}) f'(\epsilon_{t-1}) + \alpha^3 f(\epsilon_{t-3}) f'(\epsilon_{t-2}) f'(\epsilon_{t-1}) + \cdots \qquad (6.10)$$

Where, $f'(\cdot)$ denotes the derivative of $f(\cdot)$ with respect to its argument.

We emphasize that the purpose of the above analysis is to express data point $Y_t(\alpha)$, at arbitrary time t, explicitly as a function of residual terms $\epsilon_1, \epsilon_2, ..., \epsilon_t$, rather than as a function of the data itself as in the original form of the prediction of Equation (6.9). (This, to the best of our knowledge, is the only way by which the autocorrelation function $\gamma_t(\tau)$ for the data can be determined in an analytical form, because then it becomes possible to directly exploit the IID assumption about the residual sequence ϵ_t.) To achieve such a goal, the power series expansion of the nonlinear function $f(\cdot)$ has to be done around ϵ as the variable with $\alpha f(\cdot)$ as the increment, instead of expanding around $\alpha f(Y)$ with ϵ

as the increment. Two issues then arise: a) whether the series in Equation (6.10) converges; and b) if Equation 6.10 indeed converges, then how many terms should be kept in the analysis. Unfortunately, both of these issues are very difficult to address in a rigorous manner*. However, it should be reasonable to expect convergence for the series of Equation (6.10) as long as the functions $f(\cdot)$ and $f'(\cdot)$ do not grow unduly large (note that $\alpha f(\cdot)$ and $\alpha f(\cdot) + \alpha^2 f(\cdot) f'(\cdot)$ were used as the increments in the power-series expansion). Therefore, validity of the expansion in Equation (6.10) would be reasonably well held for functions of compressive nature and moderate slopes as assumed from the outset of this section.

The first and second moments of $Y_t(\alpha)$ for the model of Equation (6.7) are calculated from Equation (6.9) to give

$$E[Y_t(\alpha)] = \alpha E[f(\epsilon_{t-1})] + \alpha^2 E[f(\epsilon_{t-2}) f'(\epsilon_{t-1})] + \cdots \quad (6.11)$$

$$Var[Y_t(\alpha)] \approx Var[\epsilon_t + \alpha f(\epsilon_{t-1})] = \sigma^2 + \alpha^2 Var[f(\epsilon_{t-1})] \quad (6.12)$$

and for $\tau > 0$,

$$\begin{aligned}
&Cov[Y_t(\alpha), Y_{t+\tau}(\alpha)] \\
&\approx Cov[\epsilon_t + \alpha f(\epsilon_{t-1}) + \alpha^2 f(\epsilon_{t-2}) f'(\epsilon_{t-1}), \epsilon_{t+\tau} + \\
&\quad \alpha f(\epsilon_{t+\tau-1}) + \alpha^2 f(\epsilon_{t+\tau-2}) f'(\epsilon_{t+\tau-1})] \\
&= Cov(\epsilon_t, \epsilon_{t+\tau}) + \alpha Cov[f(\epsilon_{t-1}), \epsilon_{t+\tau}] + \alpha Cov[\epsilon_t, f(\epsilon_{t+\tau-1})] + \\
&\quad \alpha^2 Cov[f(\epsilon_{t-2}) f'(\epsilon_{t-1}), \epsilon_{t+\tau}] + \alpha^2 Cov[\epsilon_t, f(\epsilon_{t+\tau-2}) f'(\epsilon_{t+\tau-1})] + \\
&\quad \alpha^3 Cov[f(\epsilon_{t-2}) f'(\epsilon_{t-1}), f(\epsilon_{t+\tau-1})] + \\
&\quad \alpha^3 Cov[f(\epsilon_{t-1}), f(\epsilon_{t+\tau-2}) f'(\epsilon_{t+\tau-1})] + \\
&\quad \alpha^4 Cov[f(\epsilon_{t-2}) f'(\epsilon_{t-1}), f(\epsilon_{t+\tau-2}) f'(\epsilon_{t+\tau-1})] \quad (6.13)
\end{aligned}$$

Among the eight terms in Equation (6.13), the first, second, fourth, and sixth terms are zero for $\tau \geq 0$. This is due to the IID assumption for ϵ_t and to the fact that $f(\cdot)$ is a static function containing no memory. The fifth term of Equation (6.13), is non-zero only for $\tau = 1$ and $\tau = 2$. The third, seventh and the eighth terms are non-zero only for $\tau = 1$.

---

*We have found one empirical study on these issues via the method of computer simulation which provides supportive evidence. In [57], numerical results were provided for convergence of the related power series expansions for three nonlinear predictive models having the predictive functions $exp[-0.5Y^2]$, $2exp[-0.5Y^2]$, and $4 - 4exp[-0.25Y^2]$, respectively. It is interesting that all these functions are bounded in the predictive function output values, as is the case for compressive nonlinear functions which are of concern here.

A Neural Predictive HMM for Speech Recognition

Suppose, instead of approximating $Y_t(\alpha)$ by just the first three terms in Equation (6.10) above, we use higher order terms as well. Then, it can be shown that, like the above result, the covariance function will contain non-zero values also, only for small time lags up to the number of truncation terms. Since contributions of the terms in the expansion of Equation (6.10) to the degree of accuracy of the approximation are expected to decrease with higher orders of α, the covariance function would tend to attenuate very fast as time lag τ increases.

We conclude from the above analysis that prediction of a time series with a single-compressive nonlinear function does not produce long-term temporal correlations in the model's output. Note that the above analysis does not apply to a linear function for the prediction since the analysis assumes that the expansion in Equation (6.10) should converge and only the functions of compressive nature would likely satisfy this assumption.

6.3.2.3 Joint Prediction with Linear and Compressive Nonlinear Functions

Here we investigate the correlation properties of the data generated from the stationary time series model

$$Y_{t+1} = \phi Y_t + f(Y_t) + \epsilon_{t+1}, \qquad t = 1, 2, ..., T, \tag{6.14}$$

where the predictive function has a linear term additional to the nonlinear term of the model in Equation (6.7) above. Following a similar analysis to that applied to the nonlinear model, the family of models constructed for the joint model of Equation (6.14) appropriate for the ensuing perturbation analysis is

$$Y_{t+1}(\alpha) = \phi Y_t(\alpha) + \alpha f(Y_t(\alpha)) + \epsilon_{t+1}, \qquad t = 1, 2, ..., T. \tag{6.15}$$

We now decompose the stationary random process $Y_t(\alpha)$ into its stationary component processes by representing it as a power-series expansion on α

$$Y_{t+1}(\alpha) = Y_{t+1,0} + \alpha Y_{t+1,1} + \frac{1}{2!}\alpha^2 Y_{t+1,2} + \frac{1}{3!}\alpha^3 Y_{t+1,3} + \cdots. \tag{6.16}$$

In order to identify the component processes $Y_{t,i}, i = 0, 1, 2, ...$, we substitute Equation (6.16) into Equation (6.15) and approximate the nonlinear function $f(\cdot)$ by truncating its power-series expansion. This gives:

$$Y_{t+1}(\alpha) \approx \phi(Y_{t,0} + \alpha Y_{t,1} + \frac{1}{2!}\alpha^2 Y_{t,2} + \frac{1}{3!}\alpha^3 Y_{t,3})$$

$$+\alpha[f(Y_{t,0}) + f'(Y_{t,0})(\alpha Y_{t,1} + \frac{1}{2!}\alpha^2 Y_{t,2} + \frac{1}{3!}\alpha^3 Y_{t,3})] + \epsilon_{t+1}$$
$$= (\phi Y_{t,0} + \epsilon_{t+1}) + \alpha[\phi Y_{t,1} + f(Y_{t,0})] + \alpha^2[\frac{1}{2}\phi Y_{t,2} + f'(Y_{t,0})Y_{t,1}]$$
$$+\alpha^3[\frac{1}{6}\phi Y_{t,3} + \frac{1}{2}f'(Y_{t,0})Y_{t,2}] + \cdots \qquad (6.17)$$

By equating the coefficients of α^i in Equations (6.16) and (6.17), we obtain the following recursive set of relations among the component processes $Y_{t,k}$, $k = 0, 1, 2, \ldots$:

$$\begin{aligned} Y_{t+1,0} &= \phi Y_{t,0} + \epsilon_{t+1} \\ Y_{t+1,1} &= \phi Y_{t,1} + f(Y_{t,0}) \\ Y_{t+1,2} &= \phi Y_{t,2} + 2f'(Y_{t,0})Y_{t,1} \\ Y_{t+1,3} &= \phi Y_{t,3} + 3f'(Y_{t,0})Y_{t,2} \\ &\vdots \end{aligned} \qquad (6.18)$$

According to Equation (6.18), we can proceed to derive the autocovariance function for $Y_t(\alpha)$ denoted by

$$\gamma = Cov[Y_t(\alpha), Y_{t+\tau}(\alpha)] \qquad (6.19)$$

Using Equation (6.16) and truncating the expansion up to the first order, we have

$$\gamma \approx Cov[Y_{t,0} + \alpha Y_{t,1}, Y_{t+\tau,0} + \alpha Y_{t+\tau,1}] \qquad (6.20)$$

Note from Equation (6.18) that

$$\begin{aligned} Y_{t,0} + \alpha Y_{t,1} &= \phi Y_{t-1,0} + \epsilon_t + \alpha[\phi Y_{t-1,1} + f(Y_{t-1,0})] \\ &= \phi(Y_{t-1,0} + \alpha Y_{t-1,1}) + \epsilon_t + \alpha f(Y_{t-1,0}) \end{aligned} \qquad (6.21)$$

and that

$$Y_{t+\tau,0} + \alpha Y_{t+\tau,1} = \phi(Y_{t+\tau-1,0} + \alpha Y_{t+\tau-1,1}) + \epsilon_{t+\tau} + \alpha f(Y_{t+\tau-1,0}) \qquad (6.22)$$

Substituting Equations (6.21) and (6.22) into Equation (6.20) and using the stationarity property of $Y_{t,0}$ and $Y_{t,1}$ result in,

$$\begin{aligned} \gamma &= \phi^2 \gamma + \alpha^2 Cov[f(Y_{t-1,0}), f(Y_{t+\tau-1,0})] \\ &\quad + \phi\alpha Cov[Y_{t-1,0} + \alpha Y_{t-1,1}, f(Y_{t+\tau-1,0})] \\ &\quad + \phi\alpha Cov[Y_{t+\tau-1,0} + \alpha Y_{t+\tau-1,1}, f(Y_{t-1,0})] \end{aligned} \qquad (6.23)$$

A Neural Predictive HMM for Speech Recognition

Re-arranging terms and using the stationarity property of $Y_{t,0}$ and $Y_{t,1}$ again lead to,

$$\gamma = \frac{1}{(1-\phi^2)}\{\alpha^2 Cov[f(Y_{t-1,0}), f(Y_{t+\tau-1,0})] + 2\phi\alpha Cov[Y_{t,0} + \alpha Y_{t,1}, f(Y_{t+\tau,0})]\} \tag{6.24}$$

$Y_{t,0}$, the zero-th order expansion of $Y_t(\alpha)$, is a linear process and its properties are well understood. To obtain the desired form for γ, we need an explicit expression for the component covariance in Equation (6.24) involving nonlinear process $Y_{t,1}$. Repetitive use of the recursive relations in Equation (6.18) gives

$$\begin{aligned} Cov[Y_{t,1}, f(Y_{t+\tau,0})] &= Cov[\phi Y_{t-1,1} + f(Y_{t-1,0}), f(Y_{t+\tau,0})] \\ &= \phi Cov[Y_{t-1,1}, f(Y_{t+\tau,0})] + Cov[f(Y_{t-1,0}), f(Y_{t+\tau,0})] \\ &= \phi Cov[\phi Y_{t-2,1} + f(Y_{t-2,0}), f(Y_{t+\tau,0})] + Cov[f(Y_{t-1,0}), f(Y_{t+\tau,0})] \\ &\vdots \\ &= \sum_{i=0}^{\infty} \phi^i Cov[f(Y_{t-i-1,0}), f(Y_{t+\tau,0})] \end{aligned} \tag{6.25}$$

Substitution of this result into Equation (6.24) leads finally to

$$\begin{aligned} \gamma &= \frac{1}{(1-\phi^2)}\{\alpha^2 Cov[f(Y_{t-1,0}), f(Y_{t+\tau-1,0})] + 2\phi\alpha Cov[Y_{t,0}, f(Y_{t+\tau,0})] \\ &\quad + 2\phi\alpha^2 \sum_{i=0}^{\infty} \phi^i Cov[f(Y_{t-i-1,0}), f(Y_{t+\tau,0})]\} \end{aligned} \tag{6.26}$$

The first two terms in the above expression are exponentially declining as a function of time lag τ because the component processes involved are just static functions of linear processes. The remaining summation, however, would in general decay more slowly because of the many contributing terms. An empirical study of this issue was carried out through computer simulations and is presented in detail in the next subsection.

We conclude from the above result that in a model where linear and compressive nonlinear terms are jointly used for prediction, the correlation function tends to decay more slowly than in models utilizing either the linear or single-nonlinear predictive terms separately. In other words, if a model is to be constructed to represent natural data which is known to possess long-term intertime correlation, such as speech, a model of joint linear and compressive nonlinear predictive terms would be superior to (i.e. more faithful than) that of only one predictive term which is either linear or is nonlinear of a compressive type.

6.3.3 Simulation Results for the Predictive Models

Computer simulations were carried out to check the analytical results obtained in the previous subsection, where many approximations were employed to allow for the analysis to be carried out in a closed form. Using a random number generator which produces Gaussian IID residuals ϵ_t with a zero mean and unit variance, we created artificial "speech" data according to models represented by Equations (6.5), (6.7) and (6.14), respectively. The simulated data consisted of a total of 100,000 points, from which the sampled autocorrelation functions were computed for each model. The autocorrelation functions for these models were superimposed on the same plot for comparison (Figure 6.4). Parameter ϕ, interpreted as the neural network weight, is assumed to be a fixed value less than one (this guarantees stationarity of the modeled processes).

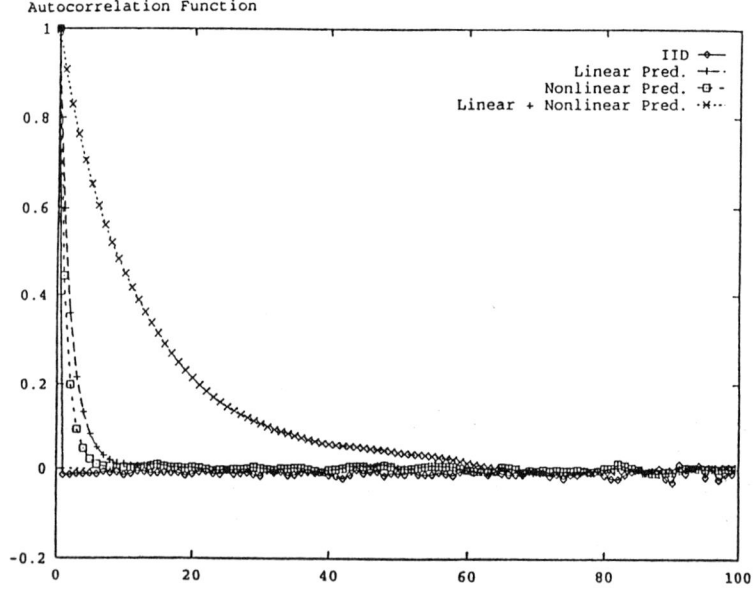

Figure 6.4 Comparison of the autocorrelation function for the: linear, nonlinear , and joint predictive models and an IID source

It is apparent that, joint use of linear and nonlinear prediction terms (i.e. The model of Equation (6.14) produces significantly stronger correlations in the simulated data than the use of separate prediction terms at any $\tau > 0$. This conforms to the analytical results obtained above. Although, the decay in the correlation function of the joint predictive model is still exponential, it happens at a much slower rate than in the other models.

Note that the correlation function of the state-dependent IID data source in the standard HMM, which is also plotted in Figure 6.4 for comparison purposes, is a delta function with zero values of correlation at any $\tau > 0$.

Figure 6.5 shows the sample autocorrelation functions for actual speech data expressed in terms of the first seven mel-frequency cepstral coefficients C1-C7 [58]. The data were taken from eight naturally spoken vowel tokens /i/ excised from the syllable /gi/. It is clear from this figure that most of these cepstral coefficients possess temporal correlation which extends over a rather long time span. For instance, the correlation for C5 extends over a time span which is significantly in excess of 20 frames ($200msec$). Referring to Figure 6.4, only the model with joint linear and compressive nonlinear prediction is able to produce significant correlation extending over more than 20 frames.

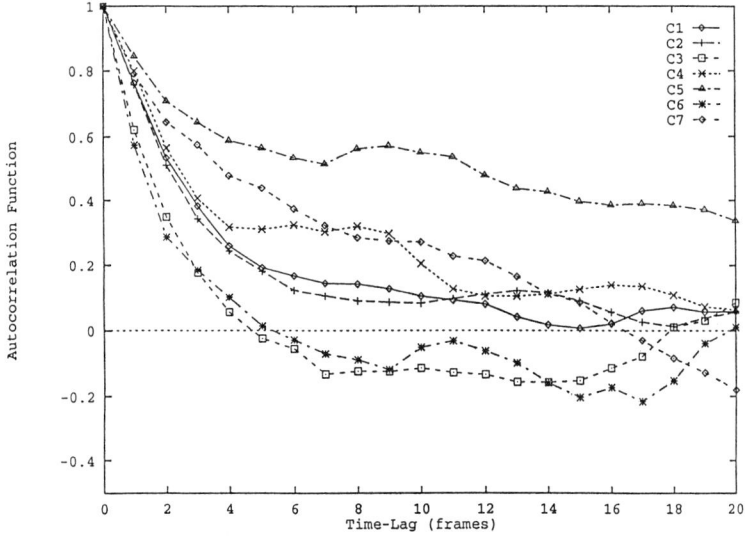

Figure 6.5 Sample autocorrelation function for the first seven mel-frequency cepstral coefficients (C1-C7).

6.3.4 Speech Recognition Experiments

The various predictive NP-HMMs discussed so far were evaluated on a speech recognition task using a database of speaker-dependent speech recorded at the University of Waterloo [37]. The vocabulary of the recognizer consisted of six Consonant-Vowel (CV) syllables, from the English E-set, where "C" encompasses six stop consonants /p/, /t/, /k/, /b/, /d/, /g/ and "V" is the vowel /i/. All the syllables were uttered with a short pause in between by native English speakers in a normal office environment. We chose this task for two reasons. First, stop confusion and E-set discrimination are known to be difficult tasks and are of fundamental significance for general speech recognition problems [44, 54]. Second, acoustic realization of stop-vowel syllables exhibits the typical nature of coarticulation and forms special sets of temporally correlated speech data. In order to faithfully represent such speech data, a model would have to be capable of handling long-term temporal correlations. The stop-E-set discrimination task therefore, allows us to perform comparative tests on the capability and effectiveness of various types of predictive models and to assess the practical value of the analytical results obtained above on actual speech recognition tasks.

Sampled speech data were obtained using a DSP Sona-Graph workstation. Data were collected by digitally sampling the speech signal at $16kHz$. A Hamming window of a 25.6-ms duration was applied every $10msec$. Within each window, a 7-dimensional vector consisting of mel-frequency cepstral coefficients was computed as raw speech data to be fed to the predictive NP-HMMs. These coefficients were appropriately scaled to accommodate the limited dynamic range in the neural network's operation. We did not use delta cepstral coefficients over time as expanded feature sets since we felt that models should be considered good only if they can generate observations which are statistically consistent with raw data. Our intention was to develop predictive HMMs as data-generator models and we believe that the advantages of using delta cepstral coefficients can be coherently embedded in the predictive mechanisms of the models.

In implementing the NP-HMMs, a four state left-to-right HMM structure was used. Each syllable in the vocabulary was represented by a series of Multi-Layer Perceptron (MLP) predictors. Each MLP predictor is associated with a state in the HMM representing a quasi-stationary segment of the speech utterance.

The MLP predictors used here consisted of a three-layered, feed-forward, and fully connected neural network. The network consisted of seven input units, one accepting each scaled cepstral coefficient. Five hidden units were employed which were either all linear, all compressively nonlinear, or a mixture of linear and compressive nonlinear units depending on the type of the predictive model considered. Seven output units were all assumed linear, each having the desired value of a corresponding predicted cepstral coefficient one frame ahead. The architecture of the MLP using a mixture of linear and compressive nonlinear units is shown in Figure 6.6.

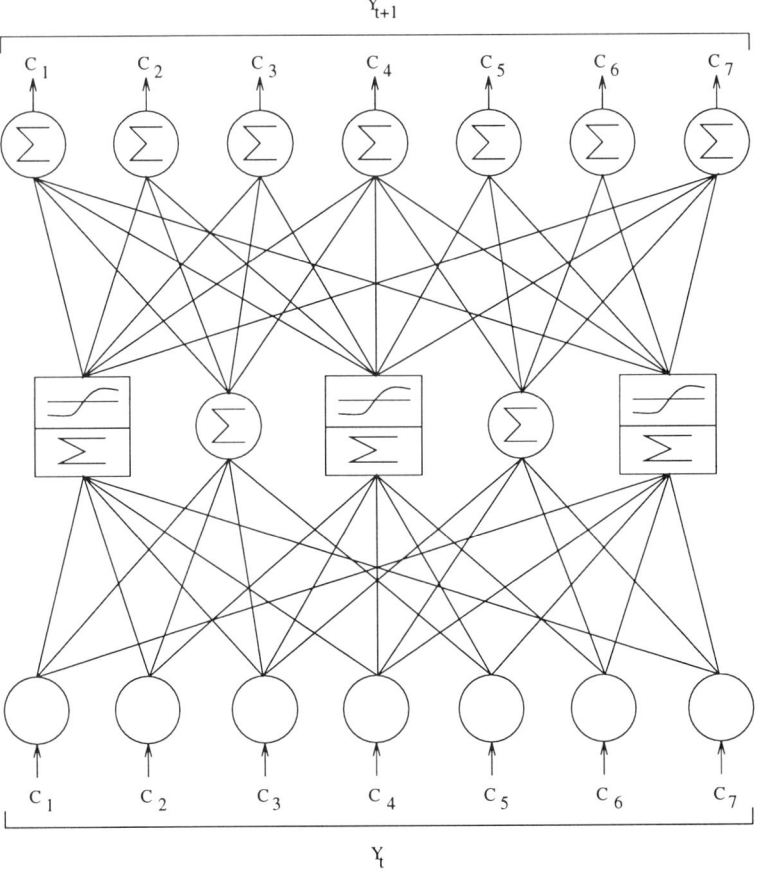

Figure 6.6 Structure of the MLP predictor using a mixture of linear and compressive nonlinear units (only partial connections are shown here to reduce clutter)

For comparison purposes, we also implemented the standard HMM, which is locally IID and is a degenerated case of the predictive HMM when the "predictive" function is fixed at a state-dependent constant (i.e. Gaussian mean). To make a fair comparison, the standard HMM was implemented with an identical structure to the NP-HMM with the mere difference of replacing data prediction with locally IID data generation. The covariance matrices in the standard HMM were assumed identity matrices in keeping with the use of the unweighted least-mean-square error function in the training and testing for the predictive HMMs.

The segmental K-means algorithm [59] was used for training the standard HMM. Dynamic programming in conjunction with Back-propagation [53] were used for training all three types of the NP-HMMs. A block diagram describing this training algorithm is shown in Figure 6.7.

The algorithm consists of two iterative steps: segmentation and estimation. The segmentation step is carried out using a standard dynamic programming procedure. The purpose of this step is to obtain the best possible segmentation of the given training utterance frames over the number of states available in a given model. This segmentation is obtained through the evaluation of the prediction performance os the various MLPs representing each state in a feed-forward pass over the entire utterance. Consequently, the best path through those states is found through a backward path.

Once the segmentation is produced, the standard Back-propagation algorithm [53] can be readily employed to estimate the MLP weight parameters associated with each segment in the estimation step. These two steps are executed iteratively until a convergence point is reached when the result of the segmentation step stays constant or when the prediction error decreases below a desired level.

When the NP-HMM is used for recognition, the best segmentation is obtained for the given test utterance on all competing models. Consequently, a feed-forward pass is performed on the MLPs of each model to calculate the total prediction error produced by each of these model for that test utterance. The utterance is then classified as belonging to the model that produces the least total prediction error.

Training and testing data were obtained from six male speakers. For each speaker, the training set consisted of eight tokens for each of the six syllables in the vocabulary. For each speaker, training was performed for an equal number of iterations for all types of predictive models considered (i.e. 15,000

A Neural Predictive HMM for Speech Recognition

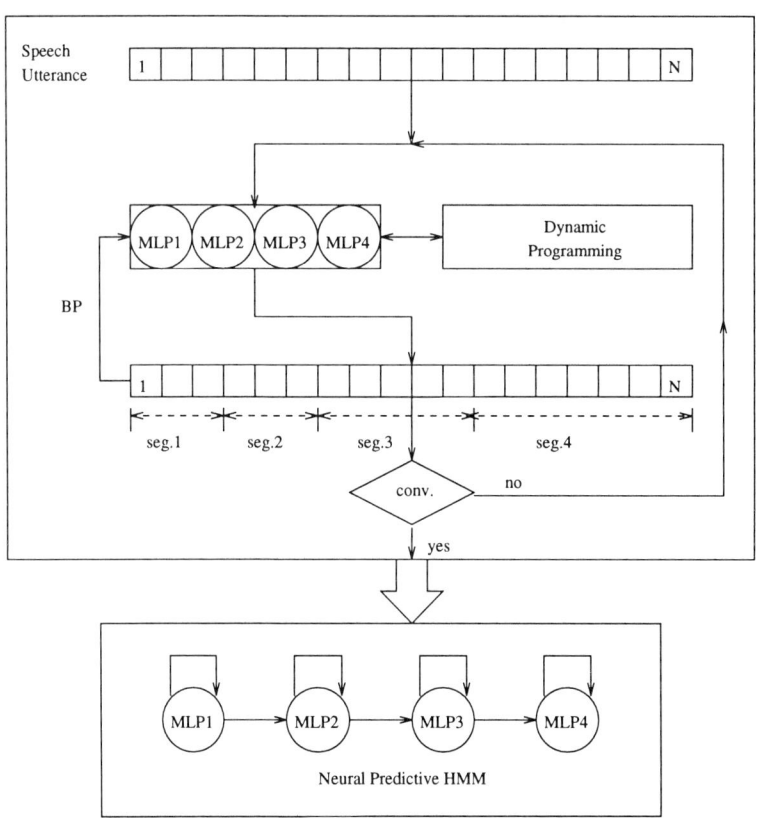

Figure 6.7 Block diagram for the algorithm used for training the NP-HMM

- 25,000 iterations) with a learning rate "η" of 0.003. The test set consisted of 14 tokens for each of the six syllables, giving a total of 84 test tokens for each speaker. Comparative recognition accuracy on the test data for the standard HMM recognizer (locally IID source with a fixed-valued "predictive" term) and for the HMM recognizers using various forms of speech-frame prediction is shown in Table 6.1. We draw particular attention to the significantly higher recognition rate (92.9%, averaged over the six speakers) obtained with mixed linear and nonlinear hidden units in the neural network architecture compared with other types of prediction (84.1%, 84.9%, and 88.1%). Further, for each individual speaker, the joint predictive model performs either better or at least as well as the best of the other three models.

Table 6.1 Comparative recognition accuracy on CV syllables

Speaker	Standard HMM	Linear Predictive HMM	Nonlinear Predictive HMM	Jointly Predictive HMM
1	85.7%	84.5%	92.8 %	97.6 %
2	91.7%	91.7%	96.4 %	100.0 %
3	80.9%	88.1%	89.3 %	89.3 %
4	80.9%	78.6%	86.9 %	89.3 %
5	94.0%	96.4%	94.0 %	96.4 %
6	71.4%	70.2%	69.0 %	84.5 %
Average	84.1%	84.9%	88.1 %	92.9%

Further experiments were performed to find out whether the linear or nonlinear models could reach the same level of performance with additional hidden units. It was found that increasing the number of hidden units beyond a certain optimal level could occasionally result in a slight improvement in performance on the training set. However, it does not yield any improvement in performance on the test set. In fact it usually results in models with poorer generalization cabilities. These results are in full agreement with the well known problem of overtraining observed whenever a more complex network structure, than needed, is used in any neural network application [60, 61].

6.3.5 Convergence Analysis of the NP-HMM

The above speech recognition results indicating the superiority of the joint NP-HMM to the purely nonlinear one, seem rather puzzling considering the fact that the sigmoid function used in the nonlinear model includes a linear range. It has been suggested that, the nonlinear model trained using the Back-propagation algorithm should be able to effectively emulate the performance of the Joint NP-HMM [63]. This could be accomplished by shrinking the values of weights feeding the hidden units that are to be operating in the linear range and enlarging the weights connecting those units to the output layer nodes. This should effectively result in a model with mixed (linear/nonlinear) hidden units, identical to that of the joint predictive model and performing joint (linear/nonlinear) prediction.

From a theoretical point of view the above scheme seems perfectly plausible, but from a practical point of view, having to optimize the weights in such a fashion so as to be operating in the limited linear range of the sigmoid nonlinearity, could lead to a slow down in convergence or to getting stuck in local minima. This, in turn, could lead to an increase in the number of iterations required for the nonlinear NP-HMM to arrive at the same performance as the joint NP-HMM.

In order to investigate this matter further, a case study was performed on speaker number 4 of Table 6.1. The purpose of this study was to establish whether the purely nonlinear model had a worse conditioning than the joint predictive model on this particular training task. The method used for evaluating the conditioning of each of these models is based on calculating the Jacobian matrices of the MLPs used in each of these two types of NP-HMMs [62]. This method is described below in more detail.

Given a training problem for an MLP, let W denote the total number of free parameters of the MLP and let N be the total number of training patterns used for training. Then, a set of W partial derivatives of the approximating function being implemented by the MLP with respect to the elements of the weight vector can be calculated for each example in the training set. This results in a matrix of partial derivatives of dimension N by W. This is called the Jacobian (J) of the MLP, where each row of J corresponds to a certain training pattern.

Many MLP training problems are intrinsically ill conditioned, resulting in a J matrix that is almost rank-deficient. The rank of J is equal to the number of non-zero singular values of J. J is said to be rank-deficient if its rank is less than min(N,W). Having a Jacobian that is rank-deficient or near rank-deficient will lead to a situation where the Back-propagation algorithm is only able to obtain partial information of the possible search directions, thus resulting in training times that are too long [62].

To investigate whether the difference in performance between the purely nonlinear NP-HMM and the joint NP-HMM was due to a difference in conditioning, the Jacobian was calculated for each of the MLPs implementing those predictive models for speaker 4 of Table 6.1.

Conditioning was evaluated by comparing the percentage of the singular values that are close to zero in each of the Jacobians of the MLPs implementing both the nonlinear and joint NP-HMMs. This is shown in Figures 6.8, and 6.9 which compare the percentage of singular values that are less than 0.01 and 0.001 respectively in both types of NP-HMMs. As could be seen from the figures, the percentage of near-zero singular values is consistently higher in the case of the nonlinear NP-HMM. This, in turn, would indicate that the Jacobians of the MLPs implementing the nonlinear model are closer to being rank deficient than those of the MLPs implementing the joint NP-HMM. Consequently slower convergence should be expected in the case of the nonlinear NP-HMM. Hence, the inferior performance exhibited by the nonlinear NP-HMM to that of the joint NP-HMM in speech recognition experiments could be attributed to its requiring more iterations to achieve the same performance due to its worse conditioning. Thus, when these two types of NP-HMMs were trained for an equal number of iterations (i.e. 15,000 in this case) it is not surprising to obtain a superior performance for the joint NP-HMM.

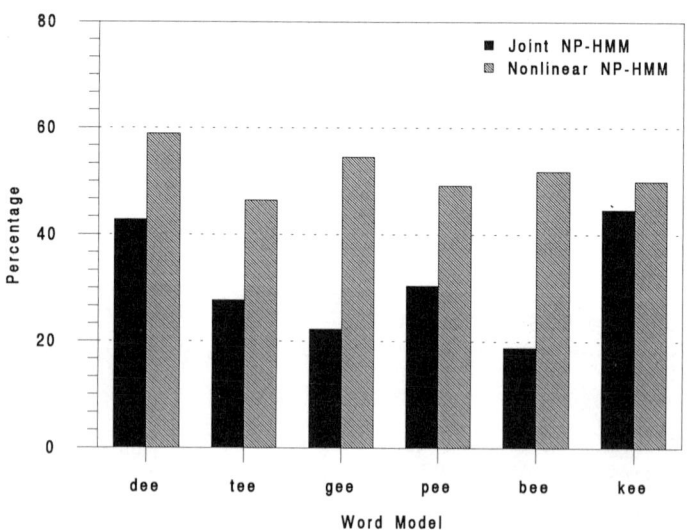

Figure 6.8 Percentage of near-zero singular values (less than 0.01) for the nonlinear and joint NP-HMM for different word models

A Neural Predictive HMM for Speech Recognition

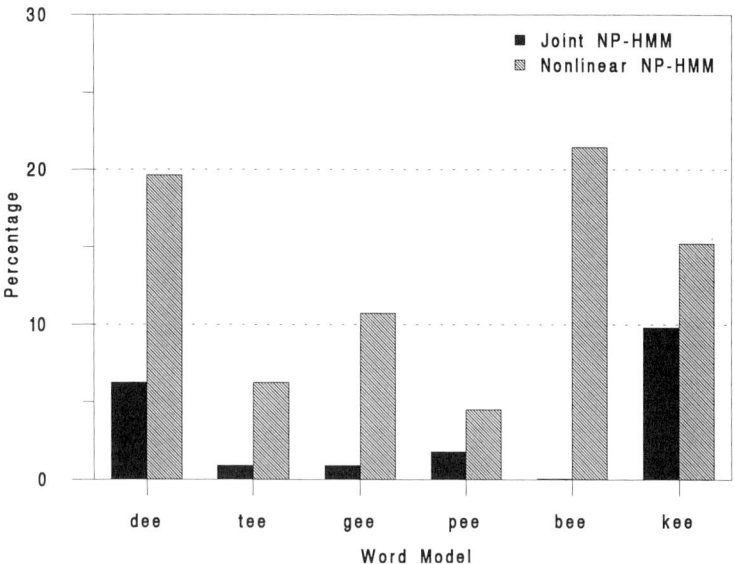

Figure 6.9 Percentage of near-zero singular values (less than 0.001) for the nonlinear and joint NP-HMM for different word models

As an alternative means of evaluating conditioning of these two types of predictive models the condition number for the Jacobian matrix (J) was calculated for each of the MLPs used in implementing both types NP-HMMs. The condition number is defined as the ratio of the largest to smallest singular values of each J [64]. The smaller the condition number, the better conditioned the system becomes. Table 6.2 compares the total condition number summed over the 4 MLPs implementing each type of predictive model. By comparing the condition numbers of the Jacobian matrices of the joint NP-HMMs to those obtained for the nonlinear NP-HMMs, it could be seen that the joint predictive model consistently exhibits lower condition numbers thus indicating, better conditioning over the pure nonlinear NP-HMMs for all word models.

The above results support the hypothesis that the improved speech recognition performance noticed when using the joint NP-HMM over the purely nonlinear NP-HMM could mainly be attributed to the improved conditioning of the training task at hand when using the joint NP-HMM. In order to check this hypothesis further, the pure nonlinear model was trained for an extra 10,000 iterations and its performance was reaccessed on the same test set as that used for the purposes of Table 6.1. The pure nonlinear model in this case was indeed able to match and even surpass the performance of the joint NP-HMM reaching a 90.05% recognition accuracy. It is concluded here therefore, that

Table 6.2 Condition numbers of the Jacobian of the nonlinear and Joint NP-HMMs for different word models

Model	Nonlinear NP-HMM	Joint NP-HMM
dee	23,240.3	3,305.0
tee	3,117.2	2,099.1
gee	4,561.6	2,421.4
pee	3,035.5	2,507.2
bee	289,513.4	24,880
kee	23,262.1	7,610.9

joint (linear/nonlinear) prediction could be equally implemented by MLPs utilizing either a mixed (linear/nonlinear) layer of hidden units or a hidden layer of purely nonlinear units with sigmoidal type nonlinearities. However, the joint NP-HMM is better conditioned and thus could reach better performance faster than the pure nonlinear NP-HMM. The theoretical justification for this disrepancy could be based on the following factors:

(i) The linear range for the joint NP-HMM is unlimited as is the case for the limited linear range for sigmoid nonlinearities found in the pure nonlinear NP-HMM. Therefore, it would be more difficult for the Back-propagation algorithm to arrive at a set of weights to utilize the limited linear range of the sigmoid nonlinearities of the pure nonlinear NP-HMM as opposed to the readily available unlimited linear range of the linear units of the joint predictive NP-HMM. This effectively means that one could get a better initialization for solving this training problem with the joint predictive model.

(ii) Occasionally the Back-propagation algorithm could get stuck in local minima. This mostly happens due to the nature of the sigmoid type nonlinearities used, which when operating in its saturated regions would result in a zero derivative term, This, in turn, would mean a near-halt in the training process if it is encountered in most of the hidden units. Having some linear units in the case of the joint NP-HMM would guard against this problem, since it would always provide a nonzero derivative, thus allowing the learning process to progress further.

The above points and the observation of the superiority of joint NP-HMMs models to pure nonlinear ones, are in full agreement with earlier observations cited by Ripley, who indicates the advantage of including "skip layer connections" which linearly link the inputs to the output layer directly [65, 66]. This architecture is very similar to the joint predictive NP-HMM used in this research, thus lending further support to the findings of this subsection.

6.4 DISCRIMINATIVE TRAINING OF THE NEURAL PREDICTIVE HMM

In spite of the clear advantages of neural-based speech predictive models, they occasionally suffer from poor discrimination. This is mainly due to the fact that each predictive word model is trained separately on its subset of the training data. This subset is not presented to any of the competing models during the training process. Hence, when a predictor gets data from a different class, it tends to produce an undefined output, which might overlap with the output of the correct predictor's model. This shortcoming is quite apparent when the prediction performance of different models turns out to be almost identical given the same data [23]. The above mode of training, in which models are independently trained, corresponds to a Maximum Likelihood (ML) training scheme [19]. A potentially good way of overcoming this problem is to utilize a corrective training scheme such as Maximum Mutual Information (MMI).

6.4.1 Maximum Mutual Information Training

Several researchers have indicated the advantages of MMI parameter estimation over that performed by ML. This was especially shown to be the case when little is known about the underlying structure of the trained model [33, 67].

In Section 6.3, we have presented a non-linear predictive HMM in which each class is represented by a series of MLP predictors, each responsible for predicting a certain segment of the speech utterance. The segmentation was modulated through a Markov chain, and training was performed on each model separately, according to an ML criterion through consecutive steps of dynamic programming and back propagation.

In this section, this work is extended by utilizing a corrective training scheme based on MMI for estimating the weights of MLP predictors. Also, a comparative performance analysis between NP-HMMs trained using ML and those trained using MMI is provided [41, 42].

In MMI training, the entire set of training data is fed to each of the models during training. The intent is to determine the weights of the model by maximizing the probability of generating the acoustic data given the correct word sequence but at the same time to minimize the probabilities of all competing models on the same word sequence.

In effect, each model is taught to facilitate data belonging to its class while inhibiting data belonging to all other classes. In fact, training of all models proceeds in parallel and weight changes in a given model depends on the current activities in competing models. The probability of a given utterance $\underline{X}$ belonging to a certain model m is governed by the relation

$$P(\underline{X} \mid m) \propto \exp\left(-\frac{1}{2} \sum_{t=1}^{T-1} \|X_{t+1} - f(X_t)\|^2\right) \qquad (6.27)$$

where $\underline{X}$ is the sequence of T speech frame vectors, $X_1 .. X_T$, constituting the given utterance to be fed sequentially to the MLP predictors of model m, and $f(X_t)$ represents the output of an MLP predictor given frame X_t as input.

Under the ML training scheme each model is trained on its subset of the training data to minimize the following energy function,

$$E = \frac{1}{N} \sum_{p=1}^{N} -\frac{1}{2} \sum_{t=1}^{T-1} \|X_{t+1}^p - f(X_t^p)\|^2 \qquad (6.28)$$

Where, p refers to a particular training sample and N is the total number of training samples.

With MMI training, we seek to maximize a rather different criterion given by,

$$M = \sum_{C=1}^{K} M_C \qquad (6.29)$$

where M_C is the mutual information evaluated over each correct model, C, and is given by

$$M_C = \frac{1}{N_c} \sum_{p=1}^{N_c} \log \left\{ \frac{P(X_c^p \mid C)}{\sum_{m=1}^{K} P(X_c^p \mid m)} \right\} \qquad (6.30)$$

Where, K is the total number of models, and N_c is the total number of training patterns (i.e. utterances) belonging to a correct model, C.

A Neural Predictive HMM for Speech Recognition

Although all models get identical pairs of inputs and desired outputs, only the correct model will be encouraged to learn while the other models are discouraged. This is due to the nature of the training criterion, M, and could be seen by taking the partial derivative of M_C with respect to P, where

$$\frac{\partial M_C}{\partial P_C} = \left(\frac{1}{P_C} - \frac{1}{\sum_{m=1}^{K} P_m} \right) \qquad (6.31)$$

$$\frac{\partial M_C}{\partial P_{m, m \neq C}} = \left(\frac{-1}{\sum_{m=1}^{K} P_m} \right) \qquad (6.32)$$

The derivative term in Equation (6.31) is always positive, thereby when multipled by the error gradient of the Back-propagation algorithm it encourages the weights of the correct model to move in a direction that improves prediction performance. In fact, this term acts as an extra adaptive learning rate for the correct model. As for the derivative term in Equation (6.32), it is always negative and as it gets multiplied with an error gradient of an incorrect model, it flips the gradient sign. Therefore, the weights of the incorrect models will be forced to move in a direction that will increase the prediction error.

6.4.2 Training Algorithms

The joint NP-HMM was implemented as explained in Subsection 6.3.4, and parameter estimation using both ML and MMI was carried out to compare their relative performance. These algorithms are described below in detail.

6.4.2.1 The ML Training Algorithm

The algorithm implementing ML estimation proceeds on each model separately as follows:

(i) Initialize the weights of all MLPs in a given model to small random values.

(ii) Segment a given training sample belonging to that model using dynamic programming in conjunction with the prediction errors from MLPs of that model (Figure 6.7).

(iii) Train each of the MLPs on its subset of the training data to decrease the prediction error on that subset.

(iv) Pick another training sample and go to step (ii) above until all the training samples belonging to this model are done.

(v) Repeat steps (ii - iv) above until either segmentation converges or the total prediction error rate is low.

(vi) Move to the next model and train it using steps (i-v) above on its subset of the training data.

6.4.2.2 The MMI Training Algorithm

The algorithm implementing MMI estimation proceeds as follows:

(i) Initialize all the weights of MLPs in all models to small random values.

(ii) Segment a given training sample using dynamic programming in conjunction with the prediction errors from MLPs on all models (Figure 6.7).

(iii) Perform a feed forward pass on all models utilizing its proper segmentation obtained in (ii) above to calculate the probability of that utterance given each of the models ($P(\underline{X}|m)$).

(iv) Train each of the MLPs in the correct model on its subset of the training data with facilitation, utilizing the probabilities from (iii) above.

(v) Train each of the MLPs in all the incorrect models on its subset of the training data with inhibition utilizing the probabilities from (iii) above.

(vi) Now, pick another training pattern and proceed to train it as in steps (ii - v) above until all the training samples from all words have been presented to all models.

(vii) Evaluate the mutual information M; if it ceases to increase significantly stop; otherwise go to step (ii) above.

6.4.3 Speech Recognition Experiments

The performance of the system, when trained with both ML and MMI, was evaluated on a speaker independent 8-vowel classification task [42]. The vowel data base used here was extracted from the ARPA TIMIT acoustic-phonetic continuous speech corpus (TIMIT).

Table 6.3 Comparative Recognition Accuracy on the TIMIT Database

ML Training	MMI Training
54.2%	65.1%

The TIMIT corpus of read speech has been designed to provide speech data for the acquisition of acoustic-phonetic knowledge and for the development and evaluation of automatic speech recognition systems. TIMIT contains a total of 6300 sentences, 10 sentences spoken by each of 630 speakers from 8 major dialect regions of the United States.

The 8 vowels used were /aa/, /ae/, /ah/, /ao/, /eh/, /ey/, /ih/, and /iy/. Vowel data from 40 speakers, representing the 8 major dialect regions of the United States were used to train the system (i.e. 3 males and 2 females from each region). This resulted in a total of 2560 training tokens. An independent test set was obtained from 24 other speakers covering again all the 8 major dialect groups (i.e. 2 males and 1 female from each region) for a total of 1346 test tokens.

The performance of the system on this task is shown in Table 6.3 with the different training criteria. The results show how MMI training was able to reduce the recognition error rate by approximately 23.8% over ML training. It is important to point out that our aim was not to obtain the best vowel classifier, but rather to compare between the two training criteria under consideration here. Thus, the overall recognition performance could have been further improved by carefully choosing optimal MLP structures to perform the required prediction tasks, and by improving the underlying HMM architecture utilized in the NP-HMM.

The difference between discriminative MMI training and nondiscriminative ML training, allowing for this substantial improvement in recognition accuracy, could be demonstrated by monitoring the variations in the class probabilities of a given training token on the various competing models as training progresses.

Figure 6.10 illustrates these probabilities during ML training for an /ae/ training token. Note again how the probabilities of both the correct model (solid line) and the incorrect models (dashed lines) settle to close values, thereby, giving rise to the possibility for regions to overlap in the prediction performance of different models, which might lead to wrong decisions when used for recognition.

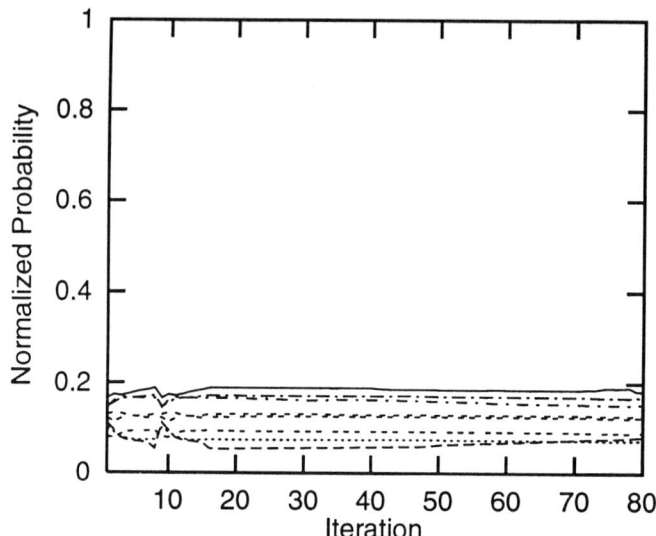

Figure 6.10 Normalized Probabilities for an 'ae' Vowel on all Models During ML Training

Figure 6.11 shows the same probabilities for models being trained with the MMI criterion on the same training token. Note again how as training progresses the probability of the correct model approaches unity while that for all incorrect models approaches zero.

The reduced overlap resulting from corrective MMI training as demonstrated by the above figures should be reflected in a higher degree of recognition accuracy since the possibility of overlap in prediction performance between competing models has been severely diminished due to the corrective nature of the MMI training criterion. This could be illustrated by examining the prediction performance on test tokens on the ML trained system versus the MMI trained one.

Figure 6.12 shows the total squared prediction error on a given test utterance of the vowel /ae/ for each of the 8 competing models in our system. As seen from the figure, prediction errors produced from all models are very small indicating good recognition performance. However, model /ae/ ranks in third place following models /eh/ and /ah/. Thus, this token will be misclassified as an /eh/.

A Neural Predictive HMM for Speech Recognition

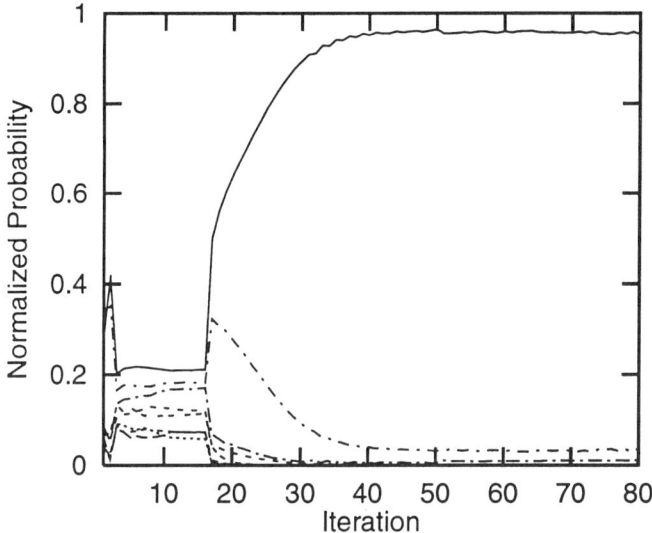

Figure 6.11 Normalized Probabilities for an 'ae' Vowel on all Models During MMI Training

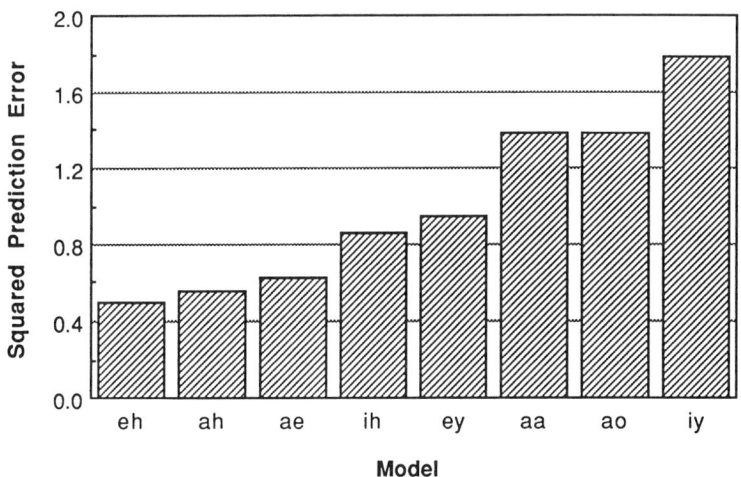

Figure 6.12 Prediction Performance for an 'ae' Test Vowel on all Models After ML Training

Figure 6.13 illustrates how MMI training avoids this situation. Here, the correct model ranks first while the /eh/ and /ah/ models have been pushed back by teaching them to predict poorly on the /eh/ vowels in the training set. This was achieved at the expense of poorer prediction performance by all models including the correct one. However, this does not matter since what is of prime concern here is not the prediction performance in itself but rather the recognition accuracy.

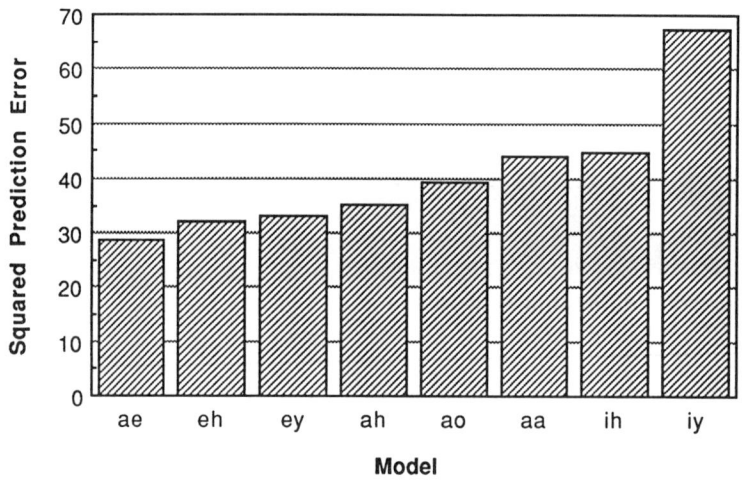

Figure 6.13 Prediction Performance for an 'ae' Test Vowel on all Models After MMI Training

Hence, the discriminative training technique based on the MMI training criterion was shown to be able to reduce areas of overlap in the predictor performance of competing word models during training. Systems trained using this corrective training approach were shown to outperform systems trained using the non-corrective ML training criterion on speech recognition tasks involving a high degree of overlap between competing classes.

One concern with this training scheme, is the computational cost involved. Roughly speaking a recognition task involving N word classes would require N times the CPU time required for ML training. This problem could be reduced by stopping the inhibition component of the MMI algorithm for models that are already highly inferior to the correct class model on the given training token. We also found that training times could be substantially reduced if corresponding MLPs in all models were initialized with the same weight values. This will give all models an equal stand at the beginning of the training process and eliminates the possibility of having the correct model being poorly initialized as compared to the rest of the models.

6.5 SPEAKER RECOGNITION USING THE NEURAL PREDICTIVE HMM

Identity verification is an integral step in controlling access to protected areas or resources. Conventional methods for achieving this goal, such as badges, keys and passwords can be stolen, duplicated or lost and hence are not very reliable. Identity recognition through speech is an emerging technology with the potential of solving all of the above problems. In addition to the ease and convenience it provides for users, it is highly accurate and can also be implemented on existing telephone networks.

In this section an automatic speaker recognition system based on a neural predictive HMM is presented [43]. The system reported here is based on fixed text speaker recognition as opposed to free text speaker recognition. In fixed text speaker recognition, the system prompts the user to utter one or a sequence of utterances which are predetermined. Free text speaker recognition systems on the other hand, accept unrestricted speech utterances. Fixed text systems are more reliable from an accuracy point of view and are sufficient for access control applications where the population of users is low and users are cooperative and consistent [68].

The speech signal contains information pertaining to both the uttered text and the characteristics of the speaker. Hence, speaker dependent speech recognition systems are well suited for the task of automatic speaker identification.

When applying speech recognition technology to the task of speaker identification, the faithfulness of the developed speech models to speech data becomes of paramount importance. This is true, since the accuracy of the identification process relies heavily on the availability of speech models exhibiting a close fit to the speech characteristics of individual speakers.

In Section 6.3 we have described an implementation of a Neural Predictive HMM (NP-HMM) for a speaker dependent speech recognition task [16]. This model was designed to capture the long term correlations existing between successive speech frames, thereby, achieving a closer fit to the actual speech data. The model utilized three-layered feed-forward ANNs as Markov-state-dependent nonlinear autoregressive-type functions in a time series model. In this section the NP-HMM is utilized for developing a fixed text speaker recognition system.

The NP-HMM was evaluated in several speaker identification tasks on the Consonant-Vowel speech database. We start by describing the details of the database used. Then we explain the speaker recognition scheme utilizing the NP-HMM. Finally, the capability of the NP-HMM speaker recognition system is demonstrated through several experiments.

6.5.1 Speech Database

The NP-HMM based speaker recognition system was evaluated using a database of speaker dependent speech recorded at the University of Waterloo. The vocabulary of the recognizer consisted of six (CV) syllables where "C" encompasses six stop consonants /p/, /t/, /k/, /b/, /d/, /g/ and "V" is the vowel /i/. All the syllables were uttered with a short pause in between by native English speakers in a normal office environment as described in Subsection 6.3.4.

Sampled speech data were obtained using a Kay-Elemetrics workstation. Data were collected by digitally sampling the speech signal at $20kHz$. A Hamming window of a $25.6msec$ duration was applied every $10msec$. Within each window, a 7-dimensional vector consisting of mel-frequency cepstral coefficients was computed as raw speech data to be fed to the predictive HMMs. These coefficients were appropriately scaled to accommodate the limited dynamic range in the ANN's operation.

Training and testing data were obtained from 6 males and 5 females for a total of 11 speakers. For each speaker, the training set consisted of eight tokens for each of the 6 syllables in the vocabulary. The test set consisted of 14 tokens for each of the six syllables, giving a total of 84 test tokens for each speaker.

6.5.2 Speaker Identification Scheme

Speaker dependent word recognition models were constructed for each speaker to characterize the parameters of his speech based on a training data set. Training proceeded as outlined in Subsection 6.3.4. The motivation behind this

A Neural Predictive HMM for Speech Recognition

speaker recognition approach is demonstrated in Figure 6.14 which depicts a typical training session for one of the speakers. Notice how the prediction performance on a test utterance, from the correct speaker, improves as training progresses following closely the improvement in prediction performance on a training utterance from the same speaker (shown by the solid line). The prediction performance on test utterances from other speakers on the other hand, does not improve beyond a certain point as shown in the figure.

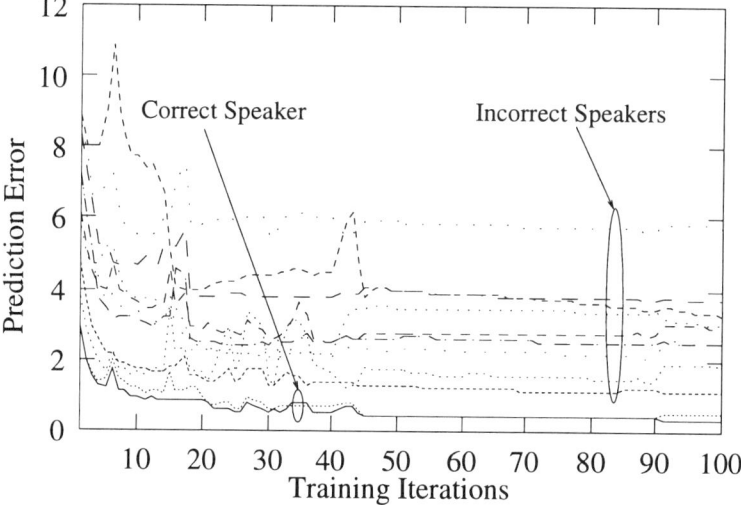

Figure 6.14 Sample training session

The speaker recognition scheme employed here is outlined in Figure 6.15. According to this scheme, a speech utterance from an anonymous speaker is first passed through a preprocessor to get the cepstral domain representation of that utterance. This representation is then passed to each of the word models developed earlier for each speaker in the database. This step is identical to what takes place when speech recognition on a given utterance is carried out. A comparison is then performed to determine the speaker model that results in the least total prediction error score for that utterance. This speaker is then chosen as the correct speaker.

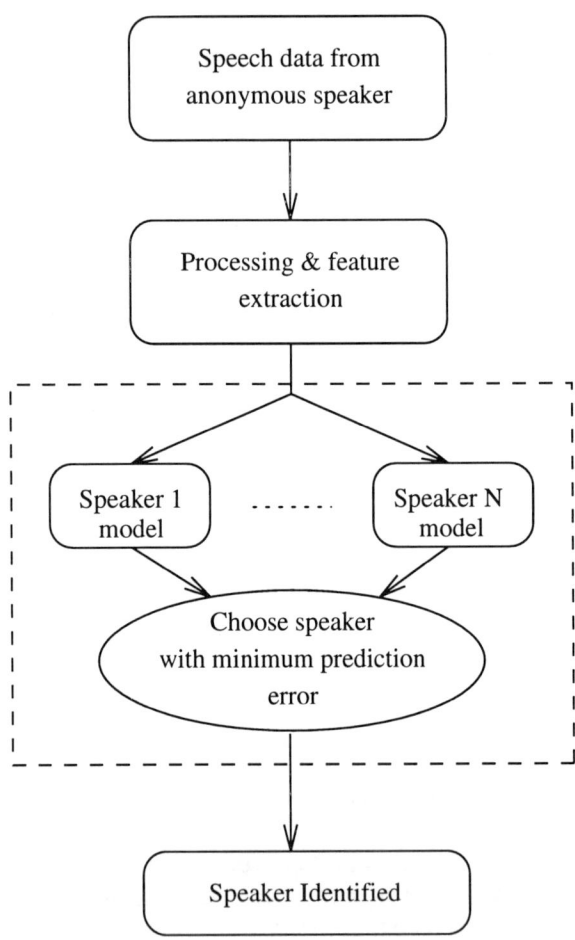

Figure 6.15 Speaker recognition scheme using the NP-HMM

6.5.3 Experiment 1: Speaker Identification

This experiment focused on the task of fixed text speaker identification. In this case, speakers had to provide a predetermined password in order to be identified correctly. The password in this case was chosen to be the syllable /di/. For the purposes of this experiment, speech data from 10 speakers (5 males and 5 females) were used to train 10 independent models for the syllable /di/. Each speaker provided a total of 22 samples of that syllable, of which 8 were used for training and 14 were used for assessing speaker recognition accuracy after all models were trained. Each model had an architecture identical to that described in Subsection 6.3.4.

The system yielded excellent performance resulting in a 99.3% speaker recognition accuracy over the 140 test samples for the same password obtained from the test token of the 10 speakers. This is an impressive result considering how short the password duration is (i.e averaging about $25 - 30 msec$).

Also, it was noticed that the single mistake made in determining a speaker's identity here was between 2 male speakers. Hence, the system could also be reliably used for applications involving sex identification.

6.5.4 Experiment 2: System Robustness

As has been pointed out in Section 6.1, speech contains a high level of variability even for the same utterance pronounced by the same speaker on two separate occasions. Therefore, a speaker recognition system should tolerate intraspeaker variations due to the speaker's mode (e.g. relaxed, stressed, or having a cold). Failing to exhibit robustness to such minor variations would render the speaker identification system totally unpractical.

In order to evaluate the robustness of our system to such variations, we again used the models trained for the 10 speakers in the previous section for the syllable /di/. However, now the performance of the system was tested with speech data representing the entire E-set database as described in Subsection 6.3.4 Hence, the test set consisted of 84 utterances for each of the 10 speakers (i.e. 14 test tokens for each of the 6 (CV) syllables). Speaker recognition accuracy over these 840 test samples was about 92.2%.

Although some degradation of performance is noticed in this case, the system still maintained a relatively good performance. The performance of the system could also be improved by retraining it with data obtained from speakers using it on a continuous basis. This will provide the system with a wider scope of data for each speaker representing his various modes of speech. Hence, the system would become incrementally more robust and would maintain good performance even in the presence of intraspeaker variability with time.

6.5.5 Experiment 3: Security Considerations

A major issue of concern when designing speaker recognition systems is security. This is especially true in cases involving access to sensitive places, such as military sites or bank accounts.

In order to make the speaker recognition system developed here more secure, we slightly modified the speaker identification scheme of Figure 6.15. The modified scheme is shown in Figure 6.16, and is identical to that of of Figure 6.15 with the exception of adding an extra stage where the prediction error score of the chosen speaker is compared to a security threshold level in order to limit the imposter acceptance rates of the system. If the prediction error score is below the given threshold, positive identification results and the speaker is accepted. If however, the error rate exceeds the security threshold, the identification result is negative and the speaker is rejected.

The security threshold is chosen to strike a balance between two opposing and very important parameters for evaluating the performance of speaker recognition systems. These are the true speaker rejection rate and the imposter acceptance rate. These two parameters should obviously be maintained at their lowest possible levels.

It was also thought that a system with several passwords would be more reliable and secure than a system employing a single password. In such a case, the system could randomly choose a word from its database and ask the user to repeat it after his entering the regular password. It would then be possible for the system to combine prediction scores from both words to be able to make a more reliable and secure decision. Such an approach would reduce the risk of imposters training to mimic true speakers, especially if the number of words in the database is large.

The security of the speaker recognition system reported here, was evaluated using the modified speaker recognition scheme of Figure 6.16. The entire CV speech database described in Section 6.3.4 was used in this experiment. This

A Neural Predictive HMM for Speech Recognition

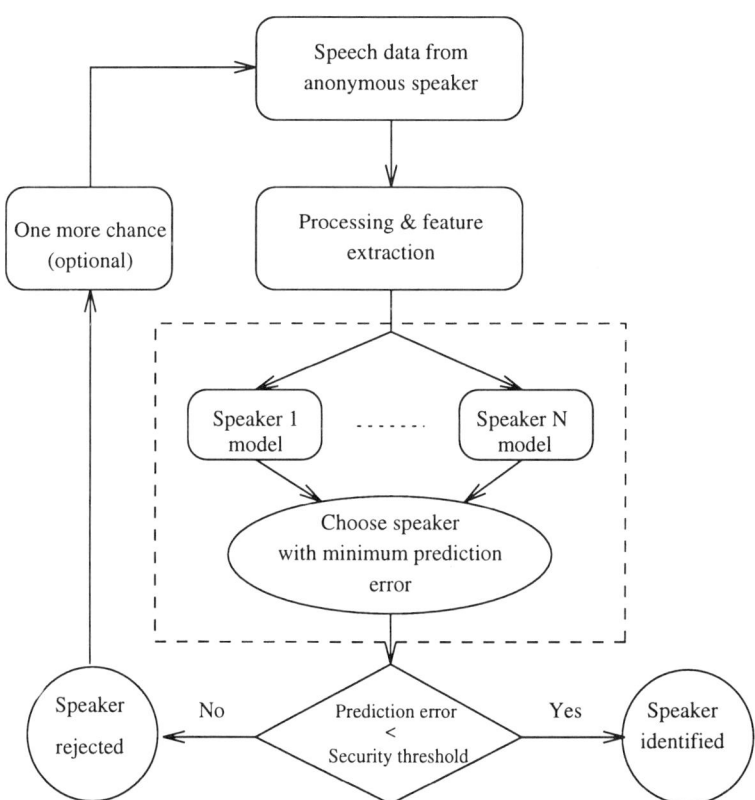

Figure 6.16 Modified speaker recognition scheme

was done for the 6 male speakers in the database only, since imposters would logically only be trying to mimic true speakers of the same sex.

Speech models were constructed for each speaker for each of the 6 CV syllables in order to assess performance on multiple passwords. Three speakers were used to train the system as before and the other three speakers were posed as imposters.

The system yielded excellent results achieving a speaker recognition rate of 100% over the 252 trials with a true speaker rejection rate of 1.6%. These 252 trials consisted of all the test tokens for the three training speakers. Also, a 0% imposter acceptance rate was achieved over the 252 trials representing the test data of imposters. Here, the security threshold was chosen to eliminate any imposters from gaining unlawful access. The 1.6% true speaker rejection rate obtained is relatively low and can be tolerated in return for maximum security, especially in military applications. Furthermore, better results could have been achieved by using multiple utterances per speaker.

6.6 CONCLUSIONS

It is concluded from this work that the signal prediction mechanism implemented by carefully structured ANNs is potentially an effective scheme for high accuracy speech recognition. In the specific task of stop-E-set recognition, use of nonlinear prediction in conjunction with linear prediction was demonstrated to be superior to either linear prediction or to nonlinear prediction carried out with a single predictive term, as well as to the standard HMM with no signal prediction. This superiority is believed to result from the higher capacity provided by this joint prediction mechanism in representing the inherent long-term correlations between successive speech frames. Analytical evaluation and computer simulations of the correlation functions for various types of simplified predictive models provide strong support for this postulation.

Theoretically, joint prediction could be carried out by MLPs either having a mixed hidden layer of linear/nonlinear units (joint NP-HMM) or having a regular hidden layer of pure nonlinear sigmoidal type units (nonlinear NP-HMM). A convergence analysis conducted to study the difference in conditioning between the nonlinear NP-HMM and the joint NP-HMM revealed that the joint NP-HMM is better conditioned than the nonlinear NP-HMM. This, in turn results, in a faster convergence in the case of the joint NP-HMM. With sufficient training however, the nonlinear NP-HMM should be able to perform as well as the joint NP-HMM. Due to this conditioning disrepency between these two types of models, it would be better to use the joint NP-HMM for this application since it would serve to reduce both training times as well as the risk of encountering local minima during the training process.

It was also demonstrated that discriminative training techniques such as MMI could be very useful in terms of reducing the possibility of accidental overlap between competing models in performing the task of speech frame prediction. This proved very useful in terms of boosting speech recognition accuracy of neural predictive HMMs.

The speaker recognition system described here was able to utilize very short speech utterances (i.e. averaging $150 - 300 msec$) to perform accurate speaker recognition. Although the results obtained here were based on a small set of speakers, they provide an indication about the validity of utilizing the NP-HMM for accurate and relatively robust speaker recognition. This speaker recognition system could be easily integrated with the speech recognition algorithm described in this chapter, thus resulting in minimal extra implementation cost. Also, it is relatively easy to enroll new speakers or update the models of existing ones.

The developed models are easy to train and it is also easy to add new words to the speech data base or new speakers to the speakers data base. The models described here are also well suited for efficient parallel VLSI implementations, for application entailing real time performance.

REFERENCES

[1] Lee, Kai-Fu, **Automatic Speech Recognition,** Kluwer Academic Publishers, 1989.

[2] Poritz, A. B., *"Hidden Markov Models: A Guided Tour,"* Proc. ICASSP, New York, 1988, pp. 7-13.

[3] Rabiner, L. R., *"A Tutorial on Hidden Markov Models and Selected Applications in Speech Recognition,"* Proc. of the IEEE, Vol. 77, No. 2, February 1989, pp. 257-285.

[4] Rabiner, L., and Juang, B., **Fundamentals of Speech Recognition,** PTR Prentice Hall, Englewood Cliffs, New Jersey, 1993.

[5] Bourlard, H. A., and Wellekens, C., *"Speech Pattern Discrimination and Multilayer Perceptrons,"* Computers, Speech & Languages, Vol. 3, 1989, pp. 1-19.

[6] Huang, W. Y., Lippmann, R. P., and Gold, B., *"A Neural Network Approach to Speech Recognition,"* Proc. ICASSP, New York, NY, April 1988, pp. 99-102.

[7] Lippmann, R. P., and Gold, B., *"Neural Classifiers Useful for Speech Recognition,"* Proc. IEEE First Int. Conf. on Neural Networks, San Diego, CA, Vol. IV, June 1987, pp. 417-425.

[8] Lippmann, R. P., *"Review of Neural Networks for Speech Recognition,"* Neural Computation, Vol. 1, 1989, pp. 1-38.

[9] Morgan, D., and Scofield, C., **Neural Networks and Speech Processing**, Kluwer Academic Publishers, Nornell, Massachusetts, 1991.

[10] Waibel, A., Hanazawa, T., Hinton, G., Shikano, K., and Lang, K., *"Phoneme Recognition Using Time-Delay Neural Networks,"* IEEE Trans. Acoustics, Speech, and Signal Processing, Vol. 37, No. 3, March 1989, pp. 328-339.

[11] Waibel, A., *" Neural Network Approaches for Speech Recognition,"* **Advances in Speech Signal Processing,** S. Furui and M. Sondhi, eds., Marcel Dekker, N. Y., 1992.

[12] Bourlard, H. A., and Wellekens, C., *"Links Between Markov Models and Multilayer Perceptrons," " Advances in Neural Information Processing Systems 1,"* D. Touretzky, ed, Morgan Kauffman, San Mateo, CA, 1991, pp. 502-510.

[13] Bourlard, H. A., *"How Connectionist Models Could Improve Markov Models for Speech Recognition,"* **Advanced Neural Computers**, R. Eckmiller Ed., Elsevier North Holland, 1990, pp. 247-254.

[14] Bridle, J. S., *"Alpha-Nets: A Recurrent 'Neural' Network Architecture with a Hidden Markov Model Interpretation,"* Speech Communication, Vol. 9, 1990, pp. 83-92.

[15] Deng, L., Hassanein, K., and Elmasry, M., *"Neural-Network Architecture for Linear and Nonlinear Predictive Hidden Markov Models: Application to Speech Recognition,"* Proc. IEEE Workshop on Neural Networks for Signal Processing, Princeton, NJ, Sept. 1991, pp. 411-421.

[16] Deng, L., Hassanein, K., and Elmasry, M., *"Analysis of the Correlation Structure for a Neural Predictive Model with Applications to Speech Recognition,"* Neural Networks, Vol. 7 (2), 1994, pp. 331-340.

[17] Franzini, M., Lee, K., and Waibel, A., *"Connectionist Viterbi Training: A New Method for Continuous Speech Recognition,"* Proc. ICASSP, Albuquerque, NM, 1990, pp. 425-428.

[18] Iso, K., and Watarnabe, T., *"Speaker-Independent Word-Recognition Using a Neural Prediction Model,"* Proc. ICASSP, 1990, pp. 441-444.

[19] Levin, E., *"Word Recognition using Hidden Control Neural Architecture,"* Proc. ICASSP, Albuquerque, NM, 1990, pp. 433-436.

[20] Niles, L., Silverman, H., Tajchman, G., and Bush, M., *"How Limited Training Data Can Allow a Neural Network Classifier to Outperform an 'Optimal' Statistical Classifier,"* Proc. ICASSP, Scotland, 1989, pp. 17-20.

[21] Niles, L., *"Modeling and Learning in Speech Recognition: The Relationship Between Stochastic Pattern Classifiers and Neural Networks,"* **Brown University Tech. Report No. LEMS-79**, 1990.

[22] Renals, S., Morgan, N., Bourlard, M., and Franco, H., *"Connectionest Probability Estimators in HMM Speech Recognition Systems,"* IEEE Trans. Speech and Audio Processing, Vol. 2(1), 1994, pp. 161-174.

[23] Tebelskis, J., and Waibel, A., *"Large Vocabulary Recognition Using Linked Predictive Neural Networks,"* Proc. ICASSP, Albuquerque, NM, 1990, pp. 437-440.

[24] Young, S. J., *"Competitive Training in Hidden Markov Models,"* Proc. ICASSP, Albuquerque, NM, 1990, pp. 681-684.

[25] Forney, G.D., *"The Viterbi Algorithm,"* Proc. of the IEEE, Vol. 61 (3), March 1973, pp. 268-278.

[26] Baum, E. L., *"An Inequality and Associated Maximization Technique in Statistical Estimation for Probabilistic Functions of Markov Processes,"* Inequalities, Vol. 3, 1972, pp. 1-8.

[27] Juang, B., and Rabiner, L. R., *"Mixture Auto-Regressive Hidden Markov Models for Speech Signals,"* IEEE Trans. Acous. Speech & Sig. Processing, Vol. ASSP-33, No. 6, Dec. 1985, pp. 1404-1413.

[28] Ney, H., and Noll, A., *"Phoneme Modeling Using Continuous Mixture Densities,"* Proc. ICASSP, 1988, pp. 437-440.

[29] Brown, P. F., *"The Acoustic-Modeling Problem in Acoustic Speech Recognition,"* **IBM Tech Report No. RC 12750**, 1987.

[30] Mariani, J., *"Recent Advances in Speech Processing,"* Proc. ICASSP, 1989, pp. 429-440.

[31] Dempster, A. P., Laird, N. M., and Rubin, D. B., *"Maximum Likelihood from Incomplete Data Via the EM Algorithm,"* Jour. Royal Stat. Soc., Vol. 39, 1977, pp. 1-38.

[32] Bahl, L. R., Brown, P. F., de Souza, P. V., and Mercer, R. L., *"A New Algorithm for the Estimation of Hidden Markov Models Parameters,"* Proc. ICASSP, New York, NY, 1988, pp. 493-497.

[33] Bahl, L. R., Brown, P. F., de Souza, P. V., and Mercer, R. L., *"Maximum Mutual Information Estimation of Hidden Markov Model Parameters for Speech Recognition,"* Proc. ICASSP, Tokyo, Japan, 1986, pp. 49-52.

[34] Merialdo, B., *"Phonetic Recognition using Hidden Markov Models and Maximum Mutual Information Training,"* ICASSP, New York, 1988, pp. 111-114.

[35] Deng, L., and Erler, K., *"Microstructural Speech Units and their HMM Representation for Discrete Utterance Speech Recognition,"* Proc. ICASSP, Toronto, Canada, May 1991, pp. 193-196.

[36] Deng, L., Lennig, M., and Mermelstein, P., *"Modeling Microsegments of Stop Consonants in a Hidden Markov Model Based Word Recognizer,"* Journal Acoust. Soc. Am., Vol. 87, 1990, pp. 2738-2747.

[37] Deng, L., and Erler, K., *"Structural Design of Hidden Markov Model Based Speech Recognizer using Multi-Valued Phonetic Features: Comparison with Segmental Speech Units,"* Journal Acoust. Soc. Am., Vol. 92 (6), 1992, pp. 3058-3067.

[38] Lippmann, R. P., *"An Introduction to Computing with Neural Nets,"* IEEE ASSP Magazine, April 1987, pp. 4-22.

[39] Kohonen, T. *"An Introduction to Neural Computing,"* Neural Networks, Vol. 1, 1988, pp. 3-16.

[40] Lang, J. K., Hinton, G. E., and Waibel, A. H., *"A Time-Delay Neural Network Architecture for Isolated Word Recognition,"*, Neural Networks, Vol. 3 (1), pp. 23-44, 1990.

[41] Hassanein, K., Deng, L., and Elmasry, M., *"Maximum Mutual Information Training of a Neural Predictive Based HMM Speech Recognition System,"* Proceedings of the Second IEEE-SP Workshop on Neural Networks for Signal Processing, Helsingor, Denmark, Aug. 1992, pp. 164-173.

[42] Hassanein, K., Deng, L., and Elmasry, M., *"Vowel Classification Using A Neural Predictive HMM: A Discriminative Training Approach,"* Proceedings of the International Conference on Acoustics, Speech and Signal Processing, Adelaide, Australia, April, 1994, Vol 2, pp. 665-668.

[43] Hassanein, K., Deng, L., and Elmasry, M., *"A Neural Predictive Hidden Markov Model for Speaker Recognition,"* Proceedings of the European Speech Communication Association Workshop on Automatic Speaker Recognition, Identification and Verification, Martigny, Switzerland, Apr. 1994, pp. 115-118.

[44] Deng, L., Kenny, P., Lennig, M., and Mermelstein, P., *"Phonemic Hidden Markov Models with Continuous Mixture Output Densities for Large Vocabulary Word Recognition,"* IEEE Trans. Signal Processing, Vol. 39, No. 7, July 1991, pp. 1677-1681.

[45] Deng, L., Kenny, P., Lennig, M., and Mermelstein, P., *"Modeling Acoustic Transitions in Speech by State-Interpolation Hidden Markov Models,"* IEEE Transactions on Signal Processing, Vol. 40 (2), 1992, pp. 265-272.

[46] Fant, G., **Acoustic Theory of Speech Production,** Mouton, The Hague, 1960.

[47] Kenny, P., Lennig, M., and Mermelstein, P., *"A Linear Predictive HMM for Vector-valued Observations with Applications to Speech Recognition,"* IEEE Trans. Acoustics, Speech, and Signal Processing, Vol. 38, No. 2, February 1990, pp. 220-225.

[48] Box, G. E. P., and Jenkins, G. M., **Time Series Analysis—Forecasting and Control,** Holden-Day, San Francisco, CA, 1976, pp. 67–72.

[49] Priestley, M., **Non-Linear and Non-Stationary Time Series Analysis**, Academic Press, London, England, 1988, pp. 140-174.

[50] Tong, H., **Non-Linear Time Series — A Dynamical System Approach**, Oxford University Press, New York, 1990.

[51] Cybenko, G., *"Approximation by Superpositions of a Sigmoidal Function,"* Mathematics of Control, Signals, and Systems, Vol. 2, June 1989, pp. 303-314.

[52] Hornik, K., Stinchcombe, M., and White, H., *"Multilayer Feed-Forward Networks are Universal Approximators,"* Neural Networks, Vol. 2, 1989, pp. 359-366.

[53] Rumelhart, D. E., Hinton, G. E., and Williams, R. J., *"Learning Internal Representations by Error Back-Propagation,"* **Parallel Distributed Processing**, Vol. 1, D.E. Rumelhart and J.L. McClelland, eds., M.I.T. Press, Cambridge, MA, 1986.

[54] Deng, L., Lennig, M., Seitz, F., and Mermelstein, P., *"Large Vocabulary Word Recognition using Context-Dependent Allophonic Hidden Markov Models,"* Computer Speech and Language, Vol. 4, No. 4, 1990, pp. 345-357.

[55] Deng, L., *"A Generalized Hidden Markov Model with State-Conditioned Trend Functions of Time for the Speech Signal,"* Signal Processing, Vol. 27 (1), 1992, pp. 65-78.

[56] Minorsky, N., **Nonlinear Oscillations**, Chapter 9, D. Van Nostrand Company Inc., Princeton, N. J., 1962.

[57] Jones, D. A., **Non-Linear Auto-Regressive Processes**, Ph.D. Thesis, University of London, 1976, pp. 132-136.

[58] Davis, S. B., and Mermelstein, P., *"Comparison of Parametric Representations for Monosyllabic Word Recognition in Continuously Spoken Sentences,"* IEEE Trans. on Acoustics, Speech and Signal Processing, Vol. 28 (4), 1980, pp. 357-365.

[59] Rabiner, L. R., Wilpon, J. G., and Juang, B. H., *"A Segmental K-means Training Procedure for Connected Word Recognition,"* AT&T Technical Journal, Vol.65, No. 3, 1986, pp. 21-31.

[60] Nowlan, J., and Hinton, E., *"Adaptive Soft Weight Tying Using Gaussian Mixtures,"* Advances in Neural Information Processing Systems 4, Morgan Kauffmann, San Mateo, CA, 1992, pp. 1-8.

[61] Weigend, A., Rumelhart, D., and Huberman, B., *"Generalization by Weight-Elimination with Applications to Forecasting,"* *" Advances in Neural Information Processing Systems 3,"* R. Lippmann, J. Moody and D. Touretzky, eds, Morgan Kauffman, San Mateo, CA, 1991.

[62] Haykin, S., **Neural Networks A Comprehensive Foundation**, IEEE Press, Macmillan College Publishing, 1994.

[63] Hinton, G. E., Personal Communication, 1994.

[64] Kailath, T., **Linear Systems,** Prentice Hall, Englewood Cliffs, New Jersey, 1980.

[65] Ripley, B. D., *"Neural Networks and Related Methods for Classification,"* Journal of the Royal Statistical Soc., Vol. 56 (3), 1994, pp. 409-456.

[66] Ripley, B. D., Personal Communication, 1994.

[67] Niles, L., Silverman, H., and Bush, M., *"Neural Networks, Maximum Mutual Information Training, and Maximum Likelihood Training,"* Proc. ICASSP, Albuquerque, NM, 1990, pp. 493-496.

[68] Naik, J., *"Speaker Verification: A Tutorial,"* IEEE Communications Magazine, January 1990, pp. 42-47.

7

MINIMUM COMPLEXITY NEURAL NETWORKS FOR CLASSIFICATION

Waleed Fakhr, Mohamed Kamel and Mohamed I. Elmasry

7.1 INTRODUCTION

Pattern classification is playing an increasingly important role in many fields of science, with applications such as medical diagnosis, hand-written character recognition, speech-to-text and text-to-speech systems, and many more. A Pattern is a set of features arranged together in a vector form, capturing information about the sources generating them. For example, a pattern in a medical diagnosis system would contain features such as patient's weight, height, age, medical symptoms, etc. A pattern classification system is a system which assigns a given pattern to one of many different classes, e.g., diseases in the medical diagnosis example.

Pattern classification may be supervised or unsupervised, and the classification system may be static or dynamic. In supervised classification, a data set of patterns and their corresponding classes is given, and the classifier is designed using this data set, which is also called a training set. In unsupervised classification, patterns are given with no labels defining their corresponding classes, and the task is to partition the space of these patterns into clusters, each containing patterns with similar features.

Static pattern classifiers are those which assign a pattern to a class or cluster based only on the information in that pattern. On the other hand, if the order at which a sequence of patterns appear carry useful information, it would not be captured by a static classifier. Instead, a dynamic classifier, which bases its decision on an ordered sequence of patterns, is needed. This is particularly the case in applications such as continuous speech recognition, and time series prediction. A static classifier may also be used for such problems, if its input is composed of a sufficient number of successive patterns. In this chapter we focus our attention on static supervised classifiers, and we also show how unsupervised classification can be used, in the supervised classification context, by employing clustering for each class of data.

Many of the neural network classifier paradigms that exist today are closely related to well known statistical pattern classification algorithms [1]. The Multilayer Perceptron "MLP" with Least Mean Square Error "LMSE" training via the Back-propagation algorithm has become the most widely used neural network classifier [1]. An MLP designer for a classification task is always faced with critical questions such as: how many layers should be used?; how many weights are required for each layer?; and what are the heuristics that should be employed during its training? Despite its popularity, there is yet no formal framework for answering these questions in a satisfactory way, for many reasons, which are discussed in the following.

Firstly, when the number of parameters is more than optimal, the MLP tends to overfit the training data, resulting in poor generalization performance [2]. Recently, many researchers have proposed various techniques to reduce the overfitting problem of the MLP. Some of these techniques add a complexity penalization term to the original classifier Mean-Square-Error "MSE" criterion, leading to a penalized maximum likelihood approach, which aims to reduce the effective number of free parameters in the MLP [2-4]. Some other techniques apply the full Bayesian inference framework * [5], or one of its approximations, where smoothing priors for the weights are used which are equivalent to the penalty terms [6,7]. An evaluation criterion, derived from the Bayesian inference framework, is computed for each trained MLP model to find the best one [8,9].

There are three drawbacks in these reported techniques. Firstly, it is not always obvious how many layers and more importantly, how many nodes in each layer are a good starting architecture for the MLP. Secondly, even though employing smoothing priors, just like performing penalized maximum likelihood estimation [10], may result in better performance, there is no guarantee that any of the weights will eventually go to zero so that it can be removed. In other words, the actual computational complexity and storage requirements of the MLP may remain unaltered. Moreover, the priors used in practice are very subjective, and a good prior in one problem may lead to bad results in another. Finally, to apply the reported evidence framework for model compar-

---

*In the Bayesian inference framework, the probability that a model M_j has generated the data D: $P(M_j|D)$ is computed for each model, and the one with the highest probability is selected.

ison, and/or for pruning weights, the normal approximation of the model-data likelihood [†] in the parameter space is essential [6-9], as well as the computation of the log-likelihood Hessian matrix of double derivatives and its inverse [6-9]. This approximation may be far from optimal, especially if the training data is small, and the model is highly nonlinear, which is the case in most practical applications. Consequently, the Hessian-based computations may be inaccurate and misleading.

For the above reasons, researchers have been investigating other alternatives which overcome the MLP drawbacks. Most prominent among these is Specht's probabilistic neural network [11-13].

The Probabilistic Neural Network "PNN", which is based on the Parzen-window technique, approximates the Probability Density Function "PDF" [‡] for each class by a Gaussian mixture or a sum of Gaussian windows (prototypes, or kernels), then uses those estimates to implement the Bayes rule for minimum-risk classification [11-13]. In contrast with the MLP [§], the PNN is a modular architecture, and each node has a well understood probabilistic interpretation. Furthermore, the PNN can be trained as a classifier in a very short time compared to the BP algorithm used to train the MLP. However, the PNN suffers from two major drawbacks. Firstly, all the training data must be stored as prototypes, making the PNN rather unattractive for VLSI implementations, and certainly allowing for potential overfitting. Secondly, the PNN lacks any form of discriminative training [¶], thus, its classification performance may be far from optimal in many applications.

The Proposed Framework

A solution to both the MLP and the PNN problems is feasible by combining the best features of both: The discriminative training of the MLP and the simple and modular architecture of the PNN. To overcome the above drawbacks, we propose the Adaptive Probabilistic Neural Network "APNN" framework [14]. In the APNN, two major issues are resolved, namely, the optimal number of

---

[†] The Likelihood is defined by $P(D|M,\theta)$, where θ is the parameter vector, and M is the model.

[‡] Given a pattern x, and a model M with a set of parameters θ, the PDF is given by $P(x|\theta, M)$. It defines the probability of that pattern being generated by this model.

[§] The MLP is not a modular architecture, where full connections between layers are usually required, particularly for multi-class systems.

[¶] Discriminative training is aimed at minimizing the classification error probability between the classes.

Gaussian prototypes for each class, and the optimal discriminative training to learn the parameters of these prototypes.

To estimate the optimal number of prototypes, we develop a Discrete Stochastic Complexity criterion "DSC" [15], which is derived from the Bayesian inference framework [6]. The DSC is evaluated for each class data separately to find the optimal number of Gaussian clusters in that class. Maximum Likelihood "ML" estimates of the model parameters are used for evaluating the DSC, therefore, an ML/DSC approach results [14,15]. The ML/DSC is optimal in the PDF estimation sense, however, it does not address the classification aspect. In many situations, however, the classification performance can be greatly enhanced by directly minimizing the Bayes error probability of the classifier, which is not done by the ML criterion. In that sense, we propose a discriminative criterion for directly minimizing the probability of error of the APNN, namely, the Maximum Mutual Information "MMI" [14-18]. The MMI is employed for two reasons. Firstly, it is directly related to a tight upper bound for the probability of error, thus its maximization leads to a direct minimization of this upper bound [16]. Secondly, maximizing the MI leads to classifier outputs which are asymptotically equal to the true class-posterior probabilities [17].

Although the APNN employs a minimal number of prototypes, all of them must be updated at each learning iteration, causing the learning to be slow. To overcome this drawback, we propose an approximation of the APNN, based on a winner-Gaussian approximation for each class PDF, which is equivalent to a nearest-neighbor approximation. Thus, it is called the Adaptive Nearest-Neighbor Classifier "ANNC" [18]. The ANNC has a similarity to the Learning Vector Quantization classifier "LVQ" [19], as both perform nearest-neighbor approximation, however, in the ANNC, the optimal number of prototypes is used, and MMI training is adopted. Unlike the LVQ, the MMI training of the ANNC does not need any heuristics, and is optimal because it directly minimizes the classification probability of error [16-18]. On the other hand, the LVQ training maximizes the divergence, which is a suboptimal criterion [17].

Further reduction of both APNN and ANNC complexities may be achieved by reducing the patterns dimensionality. This would reduce the amount of storage and the complexity of Euclidean distance computations required by both classifiers. Dimensionality reduction may be accomplished by a feature extraction transform, which extracts an optimal set of features for the probabilistic classifier. Here, we propose an adaptive feature extraction transform, which maps the I dimensional input space to a lower L dimensional space, while retaining the classification optimality of the resulting classifier. This is achieved by learning the transform coefficients at the same time as the probabilistic classifier parameters, by maximizing the mutual information of the whole classifier. This architecture is called the Adaptive Feature extraction Nearest Neighbor classifier "AFNN", where the reduced feature space is fed to the ANNC [17]. A Discrete Stochastic Complexity Criterion for classification "DSCC" is developed for the AFNN, which has its minimum for the optimal number of transform features and ANNC prototypes required for minimum error probability.

This chapter is organized as follows. In Section 7.2, we discuss both the APNN and ANNC architectures and their relation to the PNN, the Nearest Neighbor Classifier "NNC" [20] and the LVQ. In Section 7.3, the Bayesian inference framework is discussed, and the DSC for optimal PDF model selection criterion is derived. Section 7.4 presents both the ML and MMI learning equations for the proposed classifiers, where stochastic gradient ascent is used. A comparison between the APNN, the ANNC and a variety of other classifiers is shown in Section 7.5, from both performance and computational complexity aspects. The comparison is based on three experiments: A two dimensional synthetic problem, a 16 dimensional, two printed-letter recognition problem, and a 10 dimensional, 11-class vowel recognition task. The AFNN, its MMI learning, and its DSCC are presented in Section 7.6, while its performance and computational complexity is compared to other classifiers in Section 7.7. This chapter is concluded and summarized in Section 7.8.

7.2 ADAPTIVE PROBABILISTIC NEURAL NETWORKS: APNN AND ANNC

7.2.1 The PNN

The Probabilistic Neural Network "PNN", is a Parzen-window PDF estimation technique, where the true PDF of each class $P(x|C_j)$ is approximated by a sum of Gaussian windows (kernels, or prototypes) centered at the locations of the given training patterns, where x is an arbitrary pattern, and C_j denotes the j_{th} class. The PDF of the j_{th} class is estimated by:

$$P(x|\theta_j, M_j, C_j) = \frac{1}{N_j}\Sigma_{n=1}^{N_j}\Phi_n(x, \theta_n) \qquad (7.1)$$

where $P(x|\theta_j, M_j, C_j)$ is the PDF estimate for class C_j, N_j is the number of data patterns for that class, θ_j is the model parameter vector, and M_j is the model for class C_j. $\Phi_n(x, \theta_n)$ is a Gaussian window with the set of parameters θ_n which are namely the center vector and the covariance matrix. Usually the Gaussians are assumed radially symmetrical, i.e., spherical, with widths given by the standard deviations σ_n.

In Specht's original PNN [11], all Gaussians share the same width, which is determined heuristically by a leave-one-out, cross-validation technique. The PNN is used for classification by applying the Bayes rule, which assigns a pattern to the class with the highest $[P(C_j)\ P(x|\theta_j, M_j, C_j)]$. The class priors, which must sum to one, are estimated in proportion to the number of data points for each class.

There are two drawbacks in the PNN. Firstly, all training data must be stored, since there is a Gaussian window assigned to each given training pattern. Obviously this requirement may be very costly for a VLSI implementation in terms of storage area and computational time. Also, it is known by many researchers in pattern recognition that in most practical situations the PDF of each class can be very well approximated by a sum of a small number of Gaussian windows (e.g., the reduced Parzen classifier [21]), which means that the PNN usually contains potential redundancy. Moreover, as was pointed out recently, the usual kernel estimators, e.g. the PNN, are bad in the stochastic complexity sense, because to describe them we need at least as many bits as for the description of the data points themselves [22,23]. Secondly, the PNN lacks discriminative training. This may be viewed as an advantage since each class PDF network is designed completely separately, and any additional training data can be included just by adding more Gaussians to the existing ones, without having to alter the rest of the classifier. However, in many situations, especially when the training data is poor in both quantity and quality, and when the true data PDF are not well approximated by the models, discriminative training enhances classification performance.

7.2.2 The APNN

The APNN is adaptive both in architecture and parameters, where either ML, or MMI training is employed. The architecture of the APNN is adaptive in the sense that the optimal number of Gaussians is estimated through the Bayesian inference framework. Let the number of Gaussian clusters used in the APNN for the j_{th} class be K_j, which constitutes that class codebook, the PDF estimate for that class becomes:

$$P(x|\theta_j, M_j, C_j) = \frac{1}{\sum_{n=1}^{K_j} a_n} \sum_{n=1}^{K_j} a_n \, \Phi_n(x, \theta_n) \qquad (7.2)$$

where,

$$\Phi_n(x, \theta_n) = \left(\frac{2\pi^2}{\alpha_n}\right)^{-I/2} \exp\left(-\alpha_n^2 \frac{1}{2}\sum_{i=1}^{I}(x_i - m_{in})^2\right) \qquad (7.3)$$

K_j is the number of clusters for class C_j, I is the input pattern dimensionality, θ_j is the parameter vector, α_n is equal to $\frac{1}{\sigma_n}$, m_{in} is the i_{th} component of the mean vector in the n_{th} Gaussian, and $\frac{a_n}{\sum_{n=1}^{K_j} a_n}$ represents the n_{th} cluster prior probability.

A pattern x is assigned to the class with the highest posterior probability, according to the Bayes rule [20], where the j_{th} class posterior probability is given by:

$$P(C_j|x, \theta, M) = \frac{P(x|\theta_j, M_j, C_j)}{\sum_{j=1}^{J} P(x|\theta_j, M_j, C_j)} \qquad (7.4)$$

where the class priors are assumed to be equal.

The APNN is capable of forming arbitrarily smooth decision boundaries between classes, due to the Gaussian mixture modeling it incorporates. On the other hand, its learning may be slow and non-local, since all Gaussians are considered in each PDF or decision boundary computation. As an alternative, a nearest-neighbor approximation of the model in Equation (7.2) is proposed. The resulting nearest-neighbor approximation of the APNN is called the Adaptive Nearest Neighbor Classifier "ANNC".

7.2.3 The ANNC

In the ANNC, the j_{th} class PDF is approximated by:

$$P(x|\theta_j, M_j, C_j) = \frac{1}{\sum_{n=1}^{K_j} a_n} MAX_{n=1}^{K_j} a_n \, \Phi_n(x, \theta_n) \qquad (7.5)$$

where the MAX operator has replaced the Σ operator of the APNN. The MAX operator selects only the largest component in the codebook, for a given pattern, to approximate the PDF. Both the APNN and the ANNC share the same architecture, except for this difference, and that unified architecture is shown in Figure 7.1.

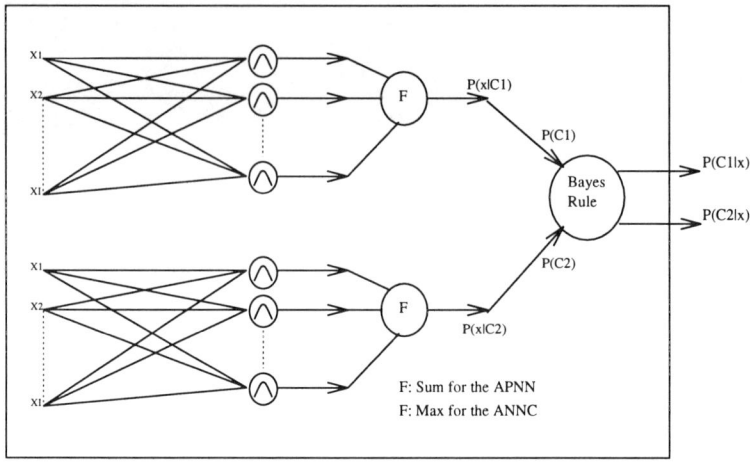

Fig(7.1): A Unified Architecture for the APNN and the ANNC

To further simplify the ANNC, we assume that all Gaussians are spherical, with the same width α_n and the same prior probability a_n. Therefore, the largest Gaussian directly represents the nearest-neighbor prototype to a given pattern. Due to this approximation, the ANNC will produce piece-wise linear approximations of the smooth decision boundaries produced by the APNN, however, the ANNC can learn potentially faster than the APNN.

The ANNC is closely related to both the nearest-neighbor classifier NNC [20] and the LVQ [19], since the three of them assign a pattern to the class with the nearest-neighbor prototype. On the other hand, in both the NNC and the LVQ, the class posterior probability is unity for the selected class and zero for the rest. In contrast, the ANNC estimates a finite posterior value to each class, according to the computed PDF of that class, similar to both the PNN and the APNN. In that sense, we expect that the ANNC performance will depend on how good is the local single-Gaussian approximation of each class PDF. In other words, it depends on how accurate the smooth decision boundary can be approximated by a piece-wise linear approximation.

The first major issue in designing either the APNN or the ANNC is to have an estimate for the optimal number of Gaussian clusters in each class. The Discrete Stochastic Complexity "DSC", derived from the Bayesian framework is used for this purpose.

7.3 BAYESIAN PDF MODEL SELECTION

Given a data set D, with patterns x_m, $m = 1, 2, .., N$, it is required to find the optimal model, among many competing ones, for estimating the PDF of the data. The Bayesian inference framework selects the model with the highest posterior probability, given the available data [6]. From the Bayes rule, the posterior probability of the j_{th} model is given by:

$$P(M_j|D) = \frac{P(M_j) \, P(D|M_j)}{P(D)} \quad (7.6)$$

where $P(M_j)$ is the prior probability of the j_{th} model, which represents our subjective belief on how plausible we thought the model is, relative to other models, before the data arrived. $P(D|M_j)$ is the likelihood of the data, given the model, which represents how likely that this set of data has been generated by this model. This term is also called the evidence for the model M_j, as revealed by the data. The normalizing factor $P(D)$ is the same for all models, thus will not make a difference in the model selection process. Finally, $P(M_j|D)$ is the posterior probability for the j_{th} model, which represents how likely we think that this model has generated the data, after the data has arrived. Eliminating any unjustified subjective belief, we choose a uniform prior over all models, thus the Bayes rule would select the model with the highest evidence $P(D|M_j)$.

An information theoretic interpretation of the evidence is given from coding theory [22,23]. It is known that the minimum number of bits required to encode a string of data, which has a probability density $P(x)$, is given by Shannon's complexity $- E_x [\log P(x)]$. In that sense, when data is modeled by a model M, we can define the number of bits required to encode the data, relative to the model, as $- \log P(D|M)$, which is the stochastic approximation of Shannon's complexity, and is thus called the Stochastic Complexity "SC" [22,23]. The SC thus gives the minimum number of bits that can be achieved to encode the data, for a specific model, and hence one has to choose the model which acquires the least SC among all competing models.

7.3.1 Asymptotic Approximations of the Stochastic Complexity

The evidence is given by:

$$P(D|M) = \int_\theta P(D|\theta, M) \, P(\theta|M) \, d\theta \quad (7.7)$$

The likelihood $P(D|\theta, M)$ is related to the model by:

$$P(D|\theta, M) = \Pi_{m=1}^{N} P(x_m|\theta, M) \qquad (7.8)$$

and the parameters prior $P(\theta|M)$ may be any proper distribution for θ. The basis of large-sample tests and confidence intervals in statistics is the property that the ML has a limiting normal distribution around the true parameter value as mean. The covariance matrix for this normal distribution is given by the inverse Hessian matrix of second derivatives, evaluated at the true parameter value. Thus, when the data set is large enough, and an ML estimate is found at θ_{ML}, the likelihood function in the parameter space may be approximated by a normal distribution, given by:

$$P(D|\theta, M) = P(D|\theta_{ML}, M) \, \exp - 1/2 \, (\theta - \theta_{ML})^T \, H^{-1}{}_{\theta_{ML}} \, (\theta - \theta_{ML}) \qquad (7.9)$$

where $H^{-1}{}_{\theta_{ML}}$ is the inverse of the Hessian matrix of second derivatives evaluated at θ_{ML}. In the above, the prior was assumed to be uniform over the parameter space. Using a uniform prior assumption, which assumes our complete ignorance about the parameter values before we receive the data, and integrating Equation (7.9) with respect to the parameters we get the evidence:

$$P(D|M) = P(D|\theta_{ML}, M) \, (2\pi)^{\frac{P}{2}} \, det^{-1/2} H_{\theta_{ML}} \qquad (7.10)$$

where $det\, H$ is the determinant of the Hessian matrix, and P is the total number of free parameters in the model. Taking the log of Equation (7.10) leads to $-SC$:

$$\log P(D|M) = \log P(D|\theta_{ML}, M) + \frac{P}{2} \log(2\pi) - \frac{1}{2} \log[det\, H_{\theta_{ML}}] \qquad (7.11)$$

Although Equation (7.11) is the asymptotic approximation of the original stochastic complexity criterion, it gives great insight into the Bayes model selection framework, since it can be written in the form:

$$\log P(D|M) = L(\theta_e) - C(\theta_e) \qquad (7.12)$$

where $L(\theta_e)$ is the log-likelihood, at the estimated parameter value, and $C(\theta_e)$ is a complexity penalizing factor, which can be called the Occam factor or the parsimony factor. Increasing the complexity of the model would make the first term increase, however, the magnitude of the second term also increases, and at a certain level of complexity, the optimal model is found, when the left hand side is maximum. The stochastic complexity SC is Equation (7.11) with a minus sign, thus the optimal model is found when the SC is minimum.

The above approximation suffers from serious drawbacks, since it is asymptotically correct, i.e., valid only when the data set is sufficiently large. We develop the DSC to overcome this drawback.

7.3.2 Discrete Stochastic Complexity Criterion "DSC"

The normal approximation of the likelihood is valid only in the asymptotic sense, i.e., when we have infinite data. Practically, it may be valid only when the number of data patterns is much larger than the number of estimated parameters. Moreover, it assumes that the likelihood distribution in the parameter space may be approximated by a single peak. On the other hand, the Minimum Description Length criterion "MDL", which is derived from the SC [25], goes even one step further, and approximates the Hessian matrix asymptotically, which is also the case for the Akaike criterion [26].

These asymptotic approximations may fail dramatically if the model is complex and the data set is small, and/or when the true distribution in the parameter space has many relevant peaks, and thus can not be approximated by a single peak. What makes matters worse is that even if the data is sufficient and the approximation is acceptable, the computation of the Hessian and its determinant is computationally very intense, particularly when the number of parameters is large. Furthermore, the Hessian is usually ill-conditioned, thus some approximations in its evaluation must be done, which may lead to misleading results.

We overcome this problem by taking advantage of the Gaussian-sum model characteristics, and treating the parameters of interest as discrete variables, with a limited set of quantized values. This allows us to perform the integration as summations over the discrete parameter space, and to compare different models in light of the evaluated discrete criterion which we call the Discrete Stochastic Complexity "DSC", without relying on the asymptotic approximation, or the Hessian evaluation. Our approach is thus suitable for small data sets, as well as large ones, unlike the asymptotic approximation approach which is suitable only for large enough data sets.

7.3.3 DSC for PDF Model Selection

The PDF approximation, using a Gaussian mixture with a model M, may be put in the form:

$$P(x_m|M,\sigma,a,m) = \frac{1}{\Sigma_{n=1}^{K} a_n}[\Sigma_{n=1}^{K} a_n \, \Phi_n(x_m, m_n, \sigma_n)] \qquad (7.13)$$

where σ is the standard deviation of the Gaussian, which is the inverse of α. The basic idea behind the discrete approximation is to quantize all the parameters, such that each comes from a set of discrete values. This proposal is also a very practical one, since whether a hardware implementation or a computer simulation program is used, the parameters always end up being quantized, as long as computations are performed digitally.

Here, the variable σ is restricted to take any value in a discrete set, selected according to the problem, and wide enough with P_σ different values to cover the significant part of σ space. In practice, the largest σ used is obtained by fitting a single Gaussian model to the data. The range between zero and σ_{max} is then divided to P_σ discrete values, with uniform steps.

The discrete variables a_n is chosen to take only the value 0 or 1. This means that the criterion gives all clusters an equal prior, and that a cluster either exists or does not exist. This may seem as a tight restriction on the model, however, it is found experimentally that the DSC can still find the optimal number of clusters, even if they have different priors, as long as the difference is within an acceptable range. This restriction simplifies the computation of the DSC drastically, and at the same time avoids finding spurious clusters, or outliers, which have small probability of appearance.

We should proceed by integrating out the means, using the same procedure of summing over all possible mean vector values. However, since each component in each Gaussian's mean may take a different value in a discrete set of P_m values, there would be $P^{(I\ K)}{}_m$ different combinations, where I is the mean vector dimensionality and K is the total number of Gaussians in the model. This, in practical situations, is a very large number, which grows exponentially with the number of Gaussians and the pattern's dimensionality, and hence another approach must be taken.

The alternative is to fix the means at their estimated ML values, and to add to the DSC criterion a penalty term equal to the cost of encoding them. This penalty term is computed by first quantizing the ML-estimated mean parameters with a proper precision d and then using a complexity term for encoding the numbers resulting from this quantization process. This complexity term depends on the dimensionality of the mean vector, the precision used, and the number of windows in the model, which enables us not only to find the optimal number of clusters, but also to test different precisions and to compare them quantitatively. The above approach was used by Rissanen in coding the means in both histogram and one-dimensional kernel estimators, where the means vector is treated as a string of integers with unknown distribution, and the universal prior for integers is used to compute the cost of encoding them [22,27]. The set of integers in this case is the mean vector components, divided by the precision, and rounded to the nearest integers. In practice, one knows the maximum magnitude value in the means vector, and an estimate for the number of discrete values required. Thus a proper precision can be selected, assuming uniform quantization steps.

The integration over the parameter space thus becomes summations over all possible values of σ and a for any given model, with the Gaussian means fixed at their ML-quantized values, and the proper penalty included. The stochastic complexity is originally given by:

$$SC = -\log \left[\int_\sigma \int_a \int_m P(\sigma|M)P(a|M)P(m|M)P(D|\sigma,a,m,M)\, d\sigma\, da\, dm \right] \quad (7.14)$$

where $P(\sigma|M)$, $P(a|M)$, and $P(m|M)$ are the prior distributions for σ, a, and m. Here, the priors for σ and a are taken to be uniform over their discrete space to reflect ignorance about their values, thus eliminating any doubt about the use of subjective priors. This gives our approach more robustness and generality over other approaches, since we do not impose any prior knowledge on the parameters distributions. Imposing nonuniform priors may lead to biased results because they may favor certain portions of the parameter space, which are not the most significant ones for the likelihood function.

7.3.4 Evaluation of the DSC

For a model with K Gaussians and P_σ discrete σ values, the properly normalized prior distributions for σ and a are:

$$P(\sigma|M) = \frac{1}{P_\sigma} \quad (7.15)$$

$$P(a|M) = \frac{1}{2^K - 1} \quad (7.16)$$

Using the discrete approximation, the above priors, and the cost for encoding the means, the DSC becomes:

$$DSC = -\log \left[\frac{1}{C_m\, P_\sigma\, [2^K - 1]} \sum_{l=1}^{P_\sigma} \sum_{a1=0}^{1} \sum_{a2=0}^{1} \cdot \sum_{a_K=0}^{1} [\Pi_{m=1}^{N} P(x_m|\sigma_l, a, M)] \right] \quad (7.17)$$

where C_m is the penalty term for encoding the means of the Gaussians, and is given by:

$$C_m = d\, \frac{(T+Q)!}{T!\, Q!} \quad (7.18)$$

where $T = \Sigma_{i=1}^{Q} |n_i|$, n_i is the nearest integer to $\frac{m_i}{d}$, d is the precision, Q is the total number of components in the means vector, and m_i is the i_{th} component in the means vector. For a dimensionality I, and a number of Gaussians K, then $Q = K \, I$.

It is to be noted that if we fix the number of discrete levels used to quantize the means, the precision d would increase when the magnitude of the mean's maximum is increased, and the penalty of encoding the means vector gets higher, as shown from Equation (7.18).

The DSC computational complexity grows exponentially with the number of Gaussians in the model. This suggests that in order to minimize the overall computational burden, we start with minimal models, and gradually increase the complexity until the DSC minimum is found.

The proposed DSC framework may thus be viewed as an adaptive clustering algorithm which has its own inherent optimality checking, and it stops only when the optimal clustering solution is found. The algorithm proposed is summarized in the following steps:

(1) Start with one Gaussian in the model, apply the ML stochastic estimation until the criterion ceases to increase, using the equations derived in the next section.

(2) From the single Gaussian model, find the suitable σ range, and the corresponding discrete σ values.

(3) Quantize the estimated mean vector with a proper precision, and compute the DSC.

(4) Increase the number of Gaussians by one, perform the ML estimation and compute the DSC.

(5) Repeat the procedure until the DSC reaches a minimum, the optimal solution is the one which corresponds to that minimum.

7.4 MAXIMUM LIKELIHOOD AND MAXIMUM MUTUAL INFORMATION TRAINING

From the DSC framework above, ML estimation is essential to determine the optimal number of Gaussian clusters per class. When this number is found, the optimal size APNN or ANNC is trained for classification using an error minimizing criterion, e.g., the MMI.

During the ML/DSC model search, the Gaussian mixture model is used, i.e., the APNN, for the following reasons:

(1) The nearest-Gaussian model, also known as the hard competitive model may suffer from the dead-unit effect, where some units receive no corrections throughout the learning process.

(2) It has been shown that the competitive model generally produces biased PDF estimates, whereas the mixture model is asymptotically unbiased.

In that sense, we start by presenting the ML and MMI learning equations for the APNN model, then the MMI equations for the ANNC model.

7.4.1 Stochastic Gradient Training in the APNN

It is interesting to first list the used criteria for the APNN learning, namely the ML and the MMI. The large sample approximations of these criteria are given by (for the ML, and MMI respectively, which need to be maximized):

$$G_L = \Sigma_{j=1}^{J} \Sigma_{m=1}^{N_j} \delta_{jm} \log P(x_m|\theta_j, M_j, C_j) \qquad (7.19)$$

$$G_{MI} = \Sigma_{j=1}^{J} \Sigma_{m=1}^{N_j} \delta_{jm} \log P(C_j|x_m, \theta, M) \qquad (7.20)$$

where δ_{jm} is 1 when the m_{th} data pattern belongs to class j, and zero otherwise, and the maximization is with respect to the parameters of the models.

Using stochastic gradient-ascent learning to maximize the above criteria, the APNN parameters, which are namely the priors, means and widths of the Gaussians, are estimated. In this stochastic approximation, parameters are updated after each pattern is presented, with its known class, until the criterion ceases to increase, or until its increase is relatively negligible. It is to be noted that the parameter learning equations for the above criteria have very similar shapes, thus in the following we present a unified learning scheme for the APNN, where by changing a criterion-dependent factor, the learning paradigm changes.

$$\Delta a_n = Z_{jm} \mu_a \frac{1}{\Sigma_{n=1}^{K_j} a_n} [\Phi_n(x_m) - \frac{\Sigma_{n=1}^{K_j} a_n \Phi_n(x_m)}{\Sigma_{n=1}^{K_j} a_n}] \qquad (7.21)$$

$$\Delta\,\alpha_n = Z_{jm}\,\mu_\alpha\,\frac{\Phi_n(x_m)a_n}{\Sigma_{n=1}^{K_j} a_n}\left[\frac{I}{\alpha_n} - \alpha_n\,\Sigma_{i=1}^{I}(x_{im} - m_{in})^2\right] \qquad (7.22)$$

$$\Delta\,m_{in} = Z_{jm}\,\mu_m\,\frac{\Phi_n(x_m)a_n}{\Sigma_{n=1}^{K_j} a_n}\,\alpha_n^{\,2}\left[x_{im} - m_{in}\right] \qquad (7.23)$$

Where $\alpha_n = \frac{1}{\sigma_n}$, μ_a, μ_α, and μ_m are the learning rates, I is the pattern dimensionality, and Z_{jm} is a factor dependent on the learning criterion, and is given by:

(1) For ML estimation, when the m_{th} pattern belongs to the j_{th} class:

$$Z_{jm} = +\,\frac{1}{P(x_m|\theta_j, M_j, C_j)} \qquad (7.24)$$

and Z_{jm} is zero otherwise.

(2) For MMI estimation, when the m_{th} pattern belongs to the j_{th} class:

$$Z_{jm} = +\,\frac{[\Sigma_{j=1}^{J} P(x_m|\theta_j, M_j, C_j)] - P(x_m|\theta_j, M_j, C_j)}{P(x_m|\theta_j, M_j, C_j)\,[\Sigma_{j=1}^{J} P(x_m|\theta_j, M_j, C_j)]} \qquad (7.25)$$

and otherwise:

$$Z_{jm} = -\,\frac{1}{\Sigma_{j=1}^{J} P(x_m|\theta_j, M_j, C_j)} \qquad (7.26)$$

7.4.2 MMI Training in the ANNC

Allowing for only the means of the Gaussians in the ANNC to adapt, and using a gradient-based stochastic approximation for optimizing the MI criterion, the learning equations are:

$$\Delta\,m_{ij} = Z_{jn}\,\mu_m\,\Phi_j(x_n)a_j\,\alpha_j^{\,2}\,(x_{in} - m_{ij}) \qquad (7.27)$$

where μ_m is the learning rates, I is the pattern dimensionality, j is the index of the winning Gaussian from the j_{th} class, and Z_{jn} is a factor dependent on the learning criterion. This factor is given by:

$$Z_{jn} = +\,\frac{[\Sigma_{j=1}^{J} P(x_n|\theta_j, M_j, C_j)] - P(x_n|\theta_j, M_j, C_j)]}{P(x_n|\theta_j, M_j, C_j)\,[\Sigma_{j=1}^{J} P(x_n|\theta_j, M_j, C_j)]} \qquad (7.28)$$

when the n_{th} pattern belongs to the j_{th} class, otherwise it is equal to:

$$Z_{jn} = -\,\frac{1}{\Sigma_{j=1}^{J} P(x_n|\theta_j, M_j, C_j)} \qquad (7.29)$$

It has to be noted that only one Gaussian per class is active for every input, thus, only the parameters of that Gaussian are allowed to adapt. In other words, for a J-class case, only J Gaussians are allowed to adapt at any given input pattern.

7.5 EXPERIMENTAL RESULTS

We show here the results of a 2 dimensional synthetic experiment, a 16 dimensional 2-letter recognition problem, and an 11-vowel recognition task with 10 dimensional patterns. These experiments are chosen to illustrate both the performance and computational complexity advantages of the proposed classifiers over the other compared ones, namely, the MLP, the Learning Vector Quantization "LVQ", the Probabilistic Neural Network "PNN", and the Nearest Neighbor Classifier "NNC".

7.5.1 Experiment 1: 2-Class Synthetic problem

This is a two-class, two dimensional problem, where the data distribution for the two classes is shown in Figure 7.2. Each class is composed of one non-

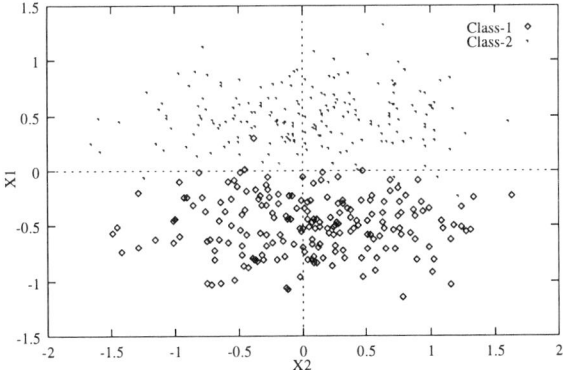

Fig(7.2): Data Distribution, Experiment 1

symmetric Gaussian cluster. The best theoretically attainable classification performance in this problem is 99%.

Here, we show a comparison between four model selection criteria versus the number of Gaussians in the PDF model (where class 1 data is used). The criteria are the DSC, the Minus-log Likelihood value denoted by "ML", the Minimum Description Length "MDL", and Akaike Information Criterion "AIC". This comparison is shown in Figures 7.3 and 7.4 for the 20 and 50 training patterns per class respectively.

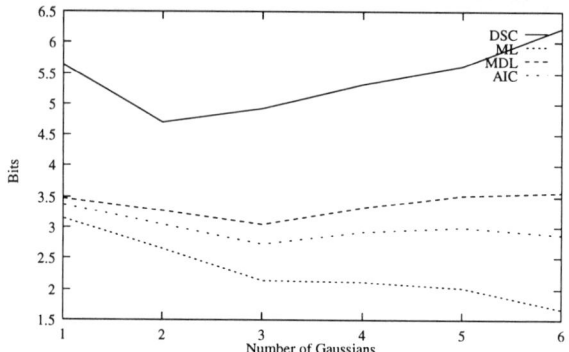

Fig(7.3): Model Selection Criteria vs Number of Gaussians
Experiment 1, 20 Pattern/Class

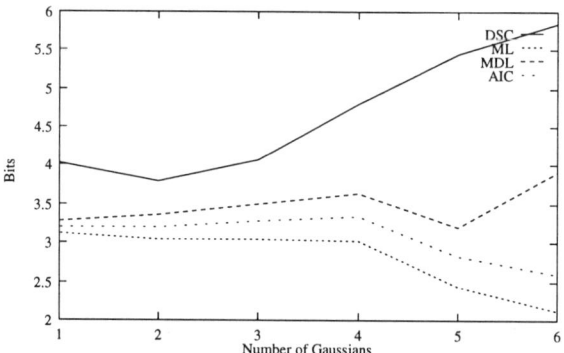

Fig(7.4): Model Selection Criteria vs Number of Gaussians
Experiment 1, 50 Patterns/Class

The first observation is that the minus-log ML value, as expected, keeps decreasing with increasing model complexity. The MDL on the other hand estimates the number of Gaussians to be 3 and 5 respectively, while the AIC also estimates 3 in the first case, but then does not find a minimum in the second case. The DSC has found the number to be 2 in both cases, because it employs a radially symmetric Gaussian model in its computations. Similar results were obtained for class 2 data.

Table 7.1 Experiment 1, Summary of Results

Classifier	%(20)	%(50)	%(100)	#(20)	#(50)	#(100)
APNN(2)	97.15	97.5	98.9	16	16	16
ANNC(2)	97.85	97.95	98	8	8	8
LVQ(2)	97.8	97.9	97.85	8	8	8
PNN	93.5	97	97.4	82	202	402
NNC	93.5	96.5	97.5	80	200	400
MLP(5)	97.4	97.8	98.5	20	20	20
MLP(10)	97.5	97.8	98.85	40	40	40
MLP(20)	97.7	97.8	98.5	80	80	80

Both the APNN and the ANNC, with two Gaussian prototypes per class, were trained by the MMI. Table 7.1 shows the comparison between the optimal size APNN and ANNC classifiers trained by the MMI criterion, with the other tested classifiers for the 20, 50, and 100 training patterns per class cases, where testing is done by a 1000 patterns/class independent set. In this table, the symbol # represents the number of parameters in the classifier, while % is the percentage correct classification on the test set. Also, a MLP(h) is a MLP with h hidden nodes, and the argument for the APNN, ANNC and the LVQ represents the number of Gaussians per class. The conclusions of this experiment are:

(1) The ANNC and the APNN have performed better or as well as the other discriminantly-trained classifiers, namely the MLP and the LVQ, while outperforming the classical classifiers, namely the PNN and the NNC, which also demand much larger number of parameters.

(2) Both the APNN and the ANNC are very compact, where only 16 and 8 parameters are stored respectively. This low complexity is matched only by the LVQ, which consistently offers less performance than the proposed classifiers.

7.5.2 Experiment 2: 2-Letter Recognition Problem

This is a 2-class problem, where the patterns are 16 dimensional extracted features for the printed letters *I* and *J* [28], and are obtained from the UCI Repository of Machine Learning Databases and Domain Theories [29]. For training,

Table 7.2 Experiment 2, Summary of Results

Classifier	%Test	#Parameters
APNN(1)	91.9	36
ANNC(1)	91.3	32
LVQ(1)	91.3	32
MLP(10)	90.4	190
MLP(20)	90.5	380
MLP(40)	90.63	760
NNC	92.88	3200
PNN	93	3200

100 patterns per class are used, while 400 per class are used for testing. The ML/DSC framework was applied for each class data separately, and has estimated that one Gaussian per class is optimal from the PDF estimation perspective. The single Gaussian per class APNN and ANNC classifiers were trained by the MMI criterion, and the classification results summary, comparing the APNN and the ANNC to other classifiers is shown in Table 7.2. The conclusions of this experiment are the following:

(1) Although the MLP has performed almost as well as the APNN and the ANNC, it requires about an order of magnitude more parameters.

(2) Surprisingly, in this experiment both the PNN and the NNC have outperformed the other classifiers, which incorporate discriminative training, however, at the expense of two orders of magnitude more parameters.

7.5.3 Experiment 3: 11-Vowel Recognition Task

In this problem we have 11 vowel classes, with 10 feature patterns, representing log (area) parameters, for the steady-state parts of the vowels: (J.A. Robinson's Data available at the UCI Repository of Machine Learning Databases and Domain Theories) [29]. For training, we used tokens from 48 speakers per class, i.e., a total of 528 patterns, while for testing, 42 different speakers were used, i.e., a total of 462 patterns.

Table 7.3 Experiment 3 (11-Vowel Task), Summary of Results

Classifier	%Test	#Parameters
APNN(31)	58.5	396
ANNC(31)	57.6	330
MLP(11)	44	242
MLP(22)	45	484
MLP(88)	51	2036
RBF(88)	48	880
RBF(528)	53	5280
NNC	56.3	5280
PNN	53	5280

The ML/DSC framework has found that 31 Gaussians are required for the whole classifier, where each class of data is treated separately. The results of the APNN and the ANNC are compared with classification results reported by J.A. Robinson in [29], as well as the NNC and the PNN, and shown in Table 7.3. From this experiment we conclude the following:

(1) Both the APNN and the ANNC have outperformed the MLP and the Radial Basis Function "RBF" classifiers by a very significant margin, as well as the NNC and the PNN.

(2) The only classifier that is close in performance to the APNN and the ANNC is the NNC, and the ratio of the number of parameters is about (1:16).

(3) This is a difficult recognition problem, since there are 11 classes, the dimensionality is relatively large, and obviously the classes heavily overlap, as the best reported result in literature was only 56.3% [29]. The proposed optimal APNN and ANNC classifiers have proved here that they are very well suited to such difficult problems, with a performance well above that of other classifiers, and a compactness which is well below other compared classifiers.

7.6 THE ADAPTIVE FEATURE EXTRACTION NEAREST NEIGHBOR CLASSIFIER "AFNN"

Probabilistic and nearest-neighbor-based classifiers can be very demanding from the computation and storage aspects depending on two factors: the number of prototypes and the input dimensionality. The LVQ [19], the APNN, and ANNC attempt to use a small number of prototypes, while retaining the classification optimality. Here, we will focus on the ANNC-type classifier, since as noted earlier, it is less complex than the APNN, in the sense that it uses only one Gaussian per class during training and operation. However, the framework proposed below may be applied to the APNN classifier with only slight modifications in the learning equations.

Although the above DSC framework may lead to significant reduction in the ANNC classifier complexity, a greater reduction can still be obtained by reducing the effective dimensionality applied to the ANNC. Not only does the large input dimensionality add complexity, but also it deters the performance, especially for small data sets in what is known as the curse of finite sample size, and the curse of dimensionality [30,31]. These effects, however, may be greatly reduced by extracting a small set of features out of the original attributes, without losing the discriminative information in the data.

Feature extraction techniques vary according to their main objective, which can be either compression or classification [20,21]. The Karhunen-Loeve Transform "KLT" is optimal in data compression applications such as transform coding, where uncorrelated features result, with only few of them containing most of the probability information required to represent the data [21]. However, the KLT may be far from optimal for classification, since it does not address the issue of discrimination between classes. This fact is demonstrated in the results shown here, as well as in [21], where other approaches were suggested instead, such as selecting transforms maximizing between-class separability, maximizing divergence, and minimizing the Bhattacharya distance. These approaches, however, suffer from computational complexity, and/or the need to make simple class distribution assumptions. In this section the Adaptive Feature extraction Nearest Neighbor classifier "AFNN" is proposed, aiming to overcome the above drawbacks.

The AFNN classifier is a hybrid architecture, and is composed of two cascaded parts. The first part is an Adaptive Feature Extractor "AFE", with a linear singular transform from the original I dimension to a smaller L dimension. The second part is an Adaptive Nearest Neighbor Classifier "ANNC", which

Minimum Complexity Neural Networks for Classification

operates on the L dimensional space, and has a codebook of K_j prototypes for the j_{th} class. Both the transform weights of the AFE and the prototype parameters of the ANNC are learned together, starting from random values, by maximizing the mutual information of the overall classifier, using a gradient ascent algorithm. The MMI learning is used here since it directly minimizes an upper bound of the classifier's probability of error as discussed in [16,17].

There are two main questions in designing an AFNN, namely: the optimal number of extracted features by the adaptive transform, and the optimal number of prototypes in the ANNC. Following the Bayesian model selection framework adopted in this research, we derive an expression for the Discrete Stochastic Complexity Criterion for classification "DSCC" in the AFNN classifier. The DSCC allows us to compare different combinations of the number of features and prototypes, and select the optimal combination among the competing ones. The optimal combination is the one which gives the least DSCC among many competing ones, which is equivalent to the model with the highest likelihood of having generated the classification data.

7.6.1 The AFNN Architecture

The AFNN is a hybrid architecture which consists of two cascaded parts: the Adaptive Feature Extractor "AFE", and the Adaptive Nearest Neighbor Classifier "ANNC" as shown in Figure 7.5 for a two-class case.

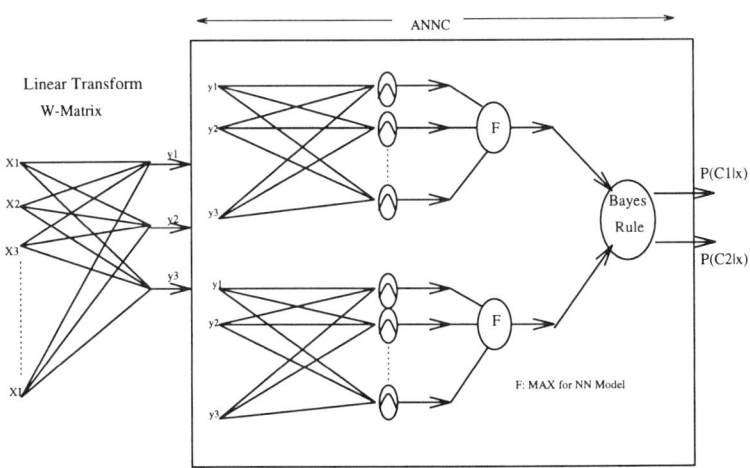

Fig(7.5): The Adaptive Feature Extraction Nearest-Neighbor Classifier, AFNN

In the AFE, the I dimensional input vector x is transformed by a linear, singular transform to the L dimensional hidden vector y. This hidden vector is the reduced dimensionality feature vector, which is used as the input pattern to the ANNC. The linear transform w obviously has $(I.L)$ weights, where the weight connecting the x_i input to the y_l hidden node or feature is denoted by w_{li}. When the n_{th} input pattern is applied, the l_{th} feature is given by:

$$y_l(n) = \Sigma_{i=1}^{I} w_{li}\, x_i(n) \tag{7.30}$$

At a given set of AFE weights, all the N training patterns of both classes are transformed to the feature space, and are viewed as the training data for the following stage, namely the ANNC.

The ANNC, as discussed earlier, is a nearest neighbor classifier, with a small number of prototypes per class which adapt their locations to optimize a certain classification criterion, for a given set of training data. The ANNC tries to find the optimal piece-wise-linear decision boundary between the classes, that optimizes the classification criterion used.

Let us focus our attention on the 2-class case, of which the multi-class case is a direct generalization, and let the codebook of the ANNC contains K_1 and K_2 prototypes for class 1 and 2 respectively. Each prototype is an L dimensional vector with each component denoted by m_{lj}, where j is the class index, and l denotes the l_{th} component in the vector. When an input pattern x is applied, a feature pattern y results, and the nearest prototypes to that pattern (in the Euclidean sense) are m_j where j is 1 and 2 for both classes respectively, which may also be called the winner prototypes.

The probabilistic approximation of the ANNC rule presented earlier, assumes that each prototype is the center or mean vector of a Gaussian window, and that for a given pattern y, each class PDF is approximated by the Gaussian centered at the winner prototype. For simplicity, we assume that all Gaussians have the same shape, and they are spherical with a standard deviation σ. In that case the winner prototype, with the least distance to the pattern, corresponds to the Gaussian with highest value, which approximates the PDF at that point. Following this formulation, the j_{th} class probability density approximation for the feature vector y, is given by:

$$P(y|\theta_j|M_j, C_j) = (2\pi\sigma^2)^{-\frac{L}{2}} \exp -\frac{1}{2\sigma^2} \Sigma_{l=1}^{L} (y_l - m_{lj})^2 \tag{7.31}$$

where j is 1 or 2, and m_{lj} is the l_{th} component of the winner prototype in the j_{th} class.

Now let us consider the mechanism by which the AFNN separates the different classes. First, the linear transform maps the data of the j_{th} class to K_j clusters in the feature space, where K_j is the number of prototypes in that class codebook. ‖ The adaptation of the transform weights, and the cluster centers, is then aimed to make the clusters of opposite classes linearly separable, so that the ANNC classifier can form proper piece-wise-linear decision boundaries to separate them.

Clearly, if the classes originally overlap, no transform can make them separable. In such cases, the AFNN would find a transform which makes the overlap as small as possible in the new feature space. The capability of the AFNN depends on the number of features L, and on the number of prototypes per class K_j, where there is always a minimal combination of both which is sufficient for the classifier to be optimal from the stochastic complexity perspective.

It is to be noted that we have chosen to use a linear transform for the feature extraction part, just like in the KLT, and other suboptimal transforms. Non-linearities can, in theory, be employed in the transform as well, e.g., by using tanh nodes, or even more than one layer of nodes. This may result in a fewer number of extracted features and prototypes, at the expense of a more complex AFE network. On the other hand, the linearity of the transform simplifies the architecture and its learning algorithm, while the performance can still be optimal by using a sufficient number of prototypes, i.e., by mapping the data to many linearly separable clusters.

In order to learn the transform and the prototype locations we maximize the mutual information for the AFNN classifier, employing the probabilistic formulation presented above.

7.6.2 MMI Training of the AFNN

For the 2-class case, the large sample approximation of the mutual information criterion for the AFNN is given by:

$$G_{MI} = (\Sigma_{n=1}^{N_1} \log P(C_1|\theta, y_n, M) + \Sigma_{n=1}^{N_2} \log P(C_2|\theta, y_n, M)) \quad (7.32)$$

where $P(C_j|\theta, y_n)$ is the posterior class probability model for the j_{th} class, and θ represents the parameters of the model (which are the mean vectors of both classes in this case, assuming the Gaussian widths and priors are fixed). From

---

‖ Assuming that all prototypes used are winners for at least one pattern.

the Bayes formula, the j_{th} class posterior probability is given by:

$$P(C_j|\theta, y_n) = \frac{P(y_n|\theta_j, M_j, C_j)}{P(y_n|\theta_1, M_1, C_1) + P(y_n|\theta_2, M_2, C_2)} \quad (7.33)$$

where the PDF model is defined in Equation (7.31).

The MI defined in Equation (7.32) is an implicit function of the original input x and the transform weights w, since the feature vector y is their linear combination. In that sense, we redefine Equation (7.32), to be an explicit function of the weights:

$$G_{MI} = \left(\sum_{n=1}^{N_1} \log P(C_1|\theta, w, x_n, M) + \sum_{n=1}^{N_2} \log P(C_2|\theta, w, x_n, M) \right) \quad (7.34)$$

where the posterior class probability is defined in Equation (7.33), but with y being substituted as a function of x and w, as given in Equation (7.34).

The MI is maximized with respect to the weights w and the prototype centers m, by a gradient-ascent algorithm. The gradient of MI with respect to the weight w_{li} can be derived by using the chain rule. For a given pattern x_n from the j_{th} class:

$$\frac{\delta G_{MI}}{\delta w_{li}} = \frac{\delta G_{MI}}{\delta P(C_j|\theta, w, x_n, M)} \frac{\delta P(C_j|\theta, w, x_n, M)}{\delta y_l} \frac{\delta y_l}{\delta w_{li}} \quad (7.35)$$

where the first term is given by:

$$\frac{\delta G_{MI}}{\delta P(C_j|\theta, w, x_n, M)} = \frac{1}{P(C_j|\theta, w, x_n, M)} = \frac{P_1 + P_2}{P_j} \quad (7.36)$$

where we denoted the PDF of the l_{th} class by P_l, which is either P_1 or P_2. The second term in the gradient is given by:

$$\frac{\delta P(C_j|\theta, w, x_n, M)}{\delta y_l} = \frac{P_1 P_2}{(P_1 + P_2)^2} \frac{1}{\sigma^2} (m_{lj} - m_{lk}) \quad (7.37)$$

where m_{lj} and m_{lk} are the l_{th} components of the winner prototype mean vectors for the j_{th} class and the opposite k_{th} class respectively. Finally, the last term is:

$$\frac{\delta y_l}{\delta w_{li}} = x_{in} \quad (7.38)$$

where x_{in} is the i_{th} component of the n_{th} input pattern. Combining the above three terms, the learning equation of the weight w_{li}, for an input x_{in} from the j_{th} class is given by:

$$\Delta w_{li} = \mu_w \frac{P_k}{(P_1 + P_2)} \frac{x_{in}}{\sigma^2} (m_{lj} - m_{lk}) \quad (7.39)$$

where μ_w is the learning gain factor, and P_k is the PDF of the opposite class.

Now we turn our attention to the learning equations for the mean vectors of the winning prototypes. For an input x_{in} from the j_{th} class, there is one winning mean vector per class, namely m_j and m_k, from the correct and incorrect class codebooks respectively. The maximum mutual information updating equations for these mean vector components are given by:

$$\Delta m_{lj} = \mu_m \frac{P_k}{(P_1 + P_2)} \frac{1}{\sigma^2} (y_l - m_{lj}) \tag{7.40}$$

$$\Delta m_{lk} = -\mu_m \frac{P_k}{(P_1 + P_2)} \frac{1}{\sigma^2} (y_l - m_{lk}) \tag{7.41}$$

where μ_m is the learning gain factor. These learning equations push the mean vector either closer to or farther from the feature vector, for the correct and incorrect class codebooks respectively. This mechanism here is identical to the one described earlier in the chapter, for the MMI training of the ANNC.

It is well known that the more complex the model is, the more flexible it becomes, and hence can form arbitrary decision boundaries. However, there is always a certain model complexity over which it starts to overfit the given training data, resulting in poor generalization to new data. On the other hand, and as a prime motivation for this work, we want to find the minimum number of features, and prototypes, which allow the classifier to form an optimal decision boundary. Thus, from both statistical and economical aspects we seek the minimal, or parsimonious AFNN classifier architecture.

To find this parsimonious architecture, we propose a Discrete Stochastic Complexity Criterion for classification "DSCC", which is derived from the Bayesian model selection framework. The DSCC for the AFNN is different from that for the APNN or the ANNC for two reasons. Firstly, in the AFNN, not only the prototypes are considered, but also the transform weights, and secondly, the DSCC is a criterion for supervised classification rather than PDF estimation. The reason the DSCC rather than the DSC is used here is that the transform in the AFNN must be designed for minimum classification error, considering different classes of data simultaneously.

7.6.3 DSCC for the AFNN

The stochastic complexity criterion for classification is given by [17]:

$$SCC = -\log P(D_c|M) \tag{7.42}$$

where M denotes the model under consideration, D_c denotes the classifier data (i.e., the training data patterns and their class labels), and $P(D_c|M)$ is the evidence for the classification data given the model, which is also the data-model likelihood of the classifier, and is given by:

$$P(D_c|M) = \int_\theta \int_w P(\theta, w|M) \, \Pi_{n=1}^N P(C_j|x_n, \theta, w, M) \, d\theta \, dw \qquad (7.43)$$

where j is the class index of the data pattern x_n, and $P(\theta, w|M)$ is the prior probability distribution of the parameters of the model, which are namely, θ of the ANNC part and w of the transform part.

The DSCC of the AFNN model, which needs to be minimal, can be approximated by:

$$DSCC = -G_{MI}(\theta_{MI}, w_{MI}) + \frac{1}{2N} \log det[\, I(w_{MI})\,]$$

$$- \frac{|W|}{2N} [\log(2\pi) - \log N] + DSCC_{annc} \qquad (7.44)$$

where $|W|$ is the number of weights, N is the number of training patterns, and:

(1) $G_{MI}(\theta_{MI}, w_{MI})$ is the mutual information of the AFNN classifier, at the MMI parameters estimates.

(2) $I(w_{MI})$ is the observed Fisher information matrix for the weights **, which is a $|W|.|W|$ symmetric matrix.

If we order the weights of the adaptive transform network in a vector of a dimension $|W|$, then the mi_{th} element of the Fisher matrix is given by:

$$f_{mi} = -\frac{\delta^2 G_{MI}}{\delta w_i \delta w_m} \qquad (7.45)$$

and for the AFNN with the linear transform, Equation (7.45) can be shown to be:

$$f_{mi} = \frac{Q}{N} \Sigma_{n=1}^N \frac{\delta G_{MI}}{\delta w_i} \frac{\delta G_{MI}}{\delta w_m} \qquad (7.46)$$

where $Q = P_1 / P_2$ for class-1 data, and its inverse for class-2 data, and the summation goes for all the training set. From Equation (7.47), the second derivatives needed for the Fisher matrix can be computed by the gradient information directly, which are already available from the learning equations, and hence simplifying the computational complexity of the DSCC significantly.

**The observed Fisher information matrix is a stochastic approximation of the expected Fisher matrix, obtained from the Hessian matrix of double derivatives

(3) The $DSCC_{annc}$ is the discrete SC for classification approximation for the ANNC part of the classifier, which is given in [17]. In this approximation, the parameters are treated as discrete (quantized) values, and the integrations over parameter space become summations over these discrete values, similar to the DSC criterion discussed earlier. Here, we used a 6-bit precision for the mean vector components, and 16 discrete values for σ, i.e., 4-bit precision. In testing the classifier results, the estimated mean parameters were also quantized by the same precision, which had no degradation effect on the performance of the classifier, as compared to non-quantized case.

(4) Finally, we assumed uniform prior probabilities for all the parameters to avoid biasing the estimates, which is the approach used in the previous chapters.

Many combinations of the number of features and prototypes for the AFNN are considered, trained by the MMI, and their DSCCs are computed. Among these competing combinations, the one with the least DSCC is selected.

7.6.4 Experimental Results

7.6.4.1 Experiment 4

This experiment is designed to demonstrate that the KLT may be incapable of performing useful dimensionality reduction in the context of classification. Here we have two classes of data, with 100 patterns each, as shown in Figure 7.6. The first class has one cluster, while the second has two clusters, and the two classes are totally separable, however, only nonlinearly, and the optimal attainable classification performance is 100%. The KLT is applied to this data, with its 2 by 2 matrix of entries given by: [0.707, 0.707, 0.707, -0.707], and the resulted transformed data is shown in Figure 7.7. From this figure we see clearly that if one tries to choose only one KLT-extracted feature, i.e., dimensionality reduction, the two classes would totally overlap. More specifically, if the X_1-axis feature is extracted, class-1 data and class-2 upper cluster data will be projected on the X_1-axis on top of each other, and a very poor classification results. Similarly, if X_2-axis feature is extracted, class-1 data and the right-most cluster of class-2 will be projected on the chosen axis on top of each other, and even poorer classification is obtained. Hence, the KLT can not be used successfully for dimensionality reduction in this case, where the basic objective is to maintain the classification performance.

276 CHAPTER 7

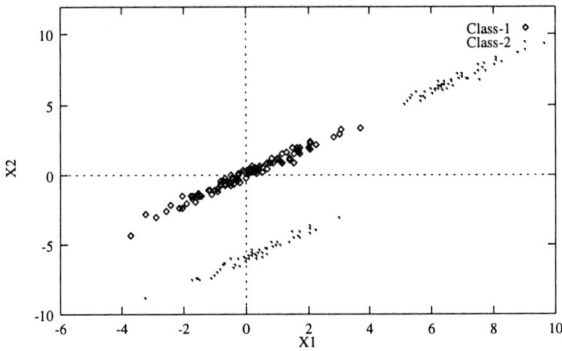
Fig(7.6): Original Data Distribution, Experiment 4

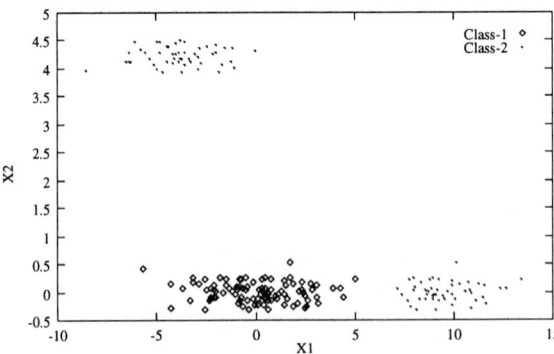
Fig(7.7): KLT-Transformed Data, Experiment 4

Minimum Complexity Neural Networks for Classification

The AFNN, on the other hand, was first applied with two extracted features (just like the KLT), and two prototypes per class. The resultant transform matrix has entries: †† [0.7788, -0.326, -0.495, 0.197], which are totally different from the KLT weights given above. The resulted transformed data is shown in Figure 7.8, where AFNN(2.2) means that 2 extracted features and 2 prototypes are used. It is clear that class-2 data is transformed into a single cluster, and that the two classes, in the new transformed space, have become linearly separable. This obviously means that there exists a line which, when this transformed data is projected on, they are still separable, i.e., only one extracted feature is sufficient. Interestingly, we can see that if the KLT is applied now for this

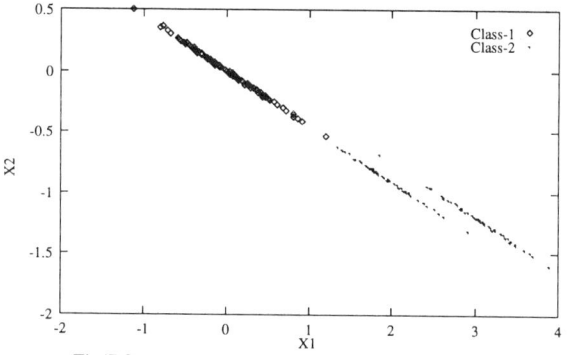

Fig(7.8): AFNN(2.2)-Transformed Data, Experiment 4

transformed data, and the data is projected along the main principal component direction, the resulting feature will result in perfect class separation.

Instead, we apply an AFNN with only one feature and two prototypes per class, since we have proof that this problem may be solved with only one extracted feature. Indeed, the AFNN obtained an optimal 100% classification solution, with weights [0.843,-0.533], where the data is projected along a line where they are separable, as shown in Figure 7.9, where each projected data point is represented by an impulse.

This experiment shows clearly that the KLT can not in general be used for classification purposes, since it is not designed to do so. Instead, we should use a classification oriented feature extraction transforms, such as the AFNN.

---
††Which are the adaptive transform weights

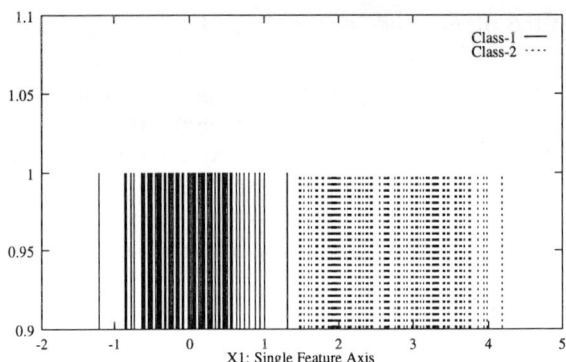

Fig(7.9): AFNN(1.2)-Transformed Data, Experiment 4

Table 7.4   Experiment 2, Summary of Results

Classifier	%Test	#Parameters
AFNN(1.1)	92	19
AFNN(1.2)	93	21
LVQ(1)	91.3	32
MLP(10)	90.4	190
MLP(20)	90.5	380
MLP(40)	90.63	760
NNC	92.88	3200
PNN	93	3200

7.6.4.2   Experiment 2 (Revisited)

This experiment, denoted earlier by experiment 2, is the printed letter recognition task, where 16 dimensional patterns for the letters *I* and *J* are used. In this problem, we also used 100 patterns per class for training and 400 per class for testing.

In this experiment, many AFNN models have close DSCC values and similar test results. The AFNN(1.1) has the least DSCC, and a good performance, however, both the AFNN(1.2) and (2.3) have slightly higher DSCC but a better performance, thus, we show here both the AFNN(1.1) and AFNN(1.2). From Table 7.4, the AFNN has outperformed all other compared classifiers from both the classification and complexity aspects. Only the PNN and the NNC have performed as well as the AFNN, however, each requires 3200 parameters, compared to only 21 parameters required by the AFNN(1.2). From this and a few other experiments shown in [17], we demonstrate how the adaptive feature extraction can be used successfully, in the context of classification. The same framework shown here for a two-class case may be applied for a multi-class architecture, where a transform is used for each class. Such a transform would aim to discriminate between that class data, and the data from all other classes, seen by this transform as the opposite class.

7.7 CONCLUSIONS

(1) We have presented a framework for designing minimum complexity probabilistic based classifiers for pattern recognition applications. The framework has two steps for the proposed classifiers, namely the APNN and the ANNC. Firstly, to find the optimal number of Gaussian clusters in each class data. This stage is performed by a ML/DSC algorithm, where a Gaussian mixture model is used for each class data. Secondly, either an APNN or ANNC classifier model is trained by the discriminative MMI training, which minimizes the probability of error.

(2) An Adaptive Feature extraction Nearest Neighbor "AFNN" classifier, for supervised classification is also proposed. The AFNN overcomes the dimensionality problems encountered in the APNN and ANNC by extracting a small set of features, optimal for classification, by MMI training.

(3) For the AFNN, a DSCC criterion for classification is developed, which enables us to compare different AFNN combinations and choose the best one. In this case, the framework is a MMI/DSCC, since MMI-trained AFNN classifiers are compared by their DSCC values.

A few conclusions may be drawn from this work:

(1) The Bayesian model selection approach is an optimal and efficient method for finding the number of clusters in a data set. This may be used for clustering, image segmentation, and classification applications.

(2) The MMI discriminative training for both APNN and ANNC improves their performances, which is in general superior to other compared classifiers.

(3) The resulting probabilistic classifiers are very compact, compared to the other considered classifiers.

(4) The AFNN is capable of mapping large dimensionality problems onto much smaller ones, and yet retaining the classification optimality.

(5) The DSCC criterion is a valuable one for selecting between AFNN classifiers, with different architectures.

REFERENCES

[1] J. Hertz, A. Krogh, and R.G. Palmer; "Introduction to the Theory of Neural Computation", Addison-Wesley, Redwood City, CA, 1991.

[2] Steven Nowlan and G.E. Hinton; "Adaptive Soft Weight Tying using Gaussian Mixtures", In J.E. Moody, S.J. Hanson and R.P. Lippmann (Eds.), Advances in Neural Information Processing Systems 4, Morgan Kaufmann, San Mateo CA 1992.

[3] Steven Nowlan and G.E. Hinton; "Simplifying Neural Networks by Soft Weight Sharing", Neural Computation, 4(4), pp.473-493, 1992.

[4] G.E. Hinton and Drew van Camp; "Keeping Neural Networks Simple by Minimizing the Description Length of the Weights", COLT-93.

[5] Radford M. Neal; "Bayesian Training of Backpropagation Networks by the Hybrid Monte Carlo Method", Technical Report CRG-TR-92-1, Connectionest Research Group, Dept. of Computer Science, U. of Toronto, April, 1992.

[6] D.J.C. Mackay; "Bayesian Methods for Adaptive Models", Ph.D. Thesis, Computation and Neural Systems, California Inst. of Technology, Pasadena, CA, 1991.

[7] D.J.C. Mackay; "A Practical Bayesian Framework for Backpropagation Networks", Neural Computation, 4(3), pp.448-472.

[8] W. Buntine, A. Weigend; "Bayesian Back-Propagation", Complex Systems, vol.5, pp(603-643), 1991.

[9] B. Hassibi and D.G. Stork; "Second Order Derivatives for Network Pruning: Optimal Brain Surgeon", Advances in Neural Information Processing Systems 4, Morgan Kaufmann, 1993.

[10] G. Wahba, C. Gu, Y. Wang and R. Chappell; "Soft Classification, a.k.a. Risk Estimation, via Penalized Log Likelihood and Smoothing Splines Analysis of Variance", To Appear in the Proceedings of the Santa Fe Workshop, D. Wolpert and A. Lapedes, Eds., Addison-Wesley, 1993.

[11] D.F. Specht; "Probabilistic Neural Networks for Classification, Mapping, or Associative Memory", Proc. IEEE International Conference on Neural Networks, Vol.I, pp.525-532, July 1988.

[12] D.F. Specht; Probabilistic Neural Networks and the Polynomial Adaline as Complementary Techniques for Classification, IEEE Transactions on Neural Networks, Vol.1, no.1, March 1990, pp(111-121).

[13] D.F. Specht; "Probabilistic Neural Networks", Neural Networks, Vol.3, pp.109-118, 1990.

[14] Waleed Fakhr, M. Kamel and M.I. Elmasry; "Minimum Complexity Adaptive Probabilistic Neural Networks for Density Estimation and Classification", Submitted to the IEEE Transactions on Neural Networks, June 1993.

[15] Waleed Fakhr, M. Kamel and M.I. Elmasry; "Unsupervised Learning by Stochastic Complexity". World Congress on Neural Networks, WCNN'93, Portland OR., July 1993.

[16] Martin E. Hellman and Josef Raviv; "Probability of Error, Equivocation, and the Chernoff Bound", IEEE Transactions on Information Theory, Vol.16, No.4, pp.368-372, July 1970.

[17] Waleed Fakhr; "Optimal Adaptive Probabilistic Neural Networks for Pattern Classification"; Ph.D. Thesis, University of Waterloo, 1993.

[18] Waleed Fakhr, M. Kamel and M.I. Elmasry; "Optimal Discriminative Training of Adaptive Nearest Neighbor Classifiers with Minimum Number of Prototypes". Submitted to the Neural Computation Journal, June 1993.

[19] T. Kohonen; "The Self-Organizing Map", Proceedings of the IEEE, Vol.78, No.9, pp.1464-1480, September 1990.

[20] R. Duda and P.E. Hart; "Pattern Classification and Scene Analysis", John Wiley & Sons, Inc., 1973.

[21] Keinosuke Fukunaga; "Introduction to Statistical Pattern Recognition", Second Edition, Academic Press, Inc. 1990.

[22] J. Rissanen; "Stochastic Complexity in Statistical Inquiry", World Scientific, 1989.

[23] J. Rissanen, Terry P. Speed and Bin Yu; "Density Estimation by Stochastic Complexity", IEEE Trans. on Information Theory, Vol.38, No.2, pp.315-323, March 1992.

[24] D.R. Cox and D.V. Hinkley; "Theoretical Statistics", Chapman and Hall, 1974.

[25] Andrew R. Barron and Thomas M. Cover; "Minimum Complexity Density Estimation", IEEE Trans. on Information Theory, Vol.37, No.4, pp.1034-1054, July 1991.

[26] H. Akaike; "Information Theory and an Extension of the Maximum Likelihood Principle", Proc. 2nd Int. Symp. Information Theory, 1972, pp.267-281.

[27] J. Rissanen; "A Universal Prior for Integers and Estimation by Minimum Description Length", The Annals of Statistics, Vol.11, No.2, pp.416-431, 1983.

[28] P.W. Frey and D.J. Slate; "Letter Recognition Using Holland Style Adaptive Classifiers", Machine Learning, Vol. 6, pp.161-182, 1991.

[29] Vowel Recognition (Deterding Data), Data Set and Results by A.J. Robinson, UCI Repository of Machine Learning Databases and Domain Theories.

[30] P.A. Devijver; "On a New Class of Bounds on Bayes Risk in Multihypothesis Pattern Recognition", IEEE Trans. on Computers, Vol.23, No.1, pp.70-80, Jan. 1974.

[31] Anil K. Jain and R. Dubes; "Feature Definition in Pattern Recognition with Small Sample Size", Pattern Recognition, Vol.10, pp.85-97., 1978.

8
A PARALLEL ANN ARCHITECTURE FOR FUZZY CLUSTERING

Dapeng Zhang, Mohamed Kamel and Mohamed I. Elmasry

8.1 INTRODUCTION

Artificial Neural Networks (ANNs) are massively parallel interconnected networks of simple (usually adaptive) nodes which are intended to interact with objects of the real world in the same way as biological nervous systems do [1]. The interest in these networks is due to the general opinion that they are able to perform some complicated and creative tasks, such as pattern recognition, similar to the way they are performed by human brains [2-3]. The implementations of these tasks by traditional computing methods have only reached relatively low performances in some limited aspects or environments. Nevertheless, as neural systems show some properties, like association, generalization, parallel searching, and adaptation to changes in the environment, which are analogous to human brain properties, they promise improved results.

However, it is a false assumption that the neural models developed in computational neuroscience, at a high level, could be directly implemented in silicon. This is because the technology, the physical devices and the circuits severely limit the performance of integrated ANNs. As a result, special ANN architecture for pattern recognition applications should be developed [4].

The complexity of ANN architecture does not stem from the complexity of its nodes but rather from the multitude of ways in which a large collection of these nodes can interact. Therefore, one important task is to build highly parallel, regular and modular ANN architectures based on data matrices that make them attractive for VLSI techniques [5-8]. Some typical ANN architectures for pattern classification and recognition are:

Feedforward Network: The basic structure is the matrix multiplication, where the input vector is multiplied by the weights stored in the network, the products are summed up and evaluated by a nonlinearity function [9];

Feedback Network: The output of the network is connected to the input to yield a more complex behavior of the network, which can be characterized by minimizing an "energy" or cost function [10];

Closed Loop Antagonistic Network: It evaluates the match and the mismatch between the input information and the stored patterns to realize networks which are more dynamic [11].

In this chapter, we present a parallel neural network architecture for fuzzy clustering application, called Fuzzy Clustering Neural Network (FCNN) architecture. The organization of the chapter is arranged as follows: In Section 8.2, we introduce some basic concepts about fuzzy clustering and neural networks. The fuzzy competitive learning algorithm and the corresponding FCNN architecture are discussed in Section 8.3 and 8.4, respectively. In Section 8.5, we investigate three types of processing cells used in the FCNN architecture. Compared with the Fuzzy C-Mean (FCM) algorithm, Section 8.6 analyzes the performance of the FCNN architecture in terms of simulation results and hardware complexity. In Section 8.7, we summarize the conclusions of the chapter.

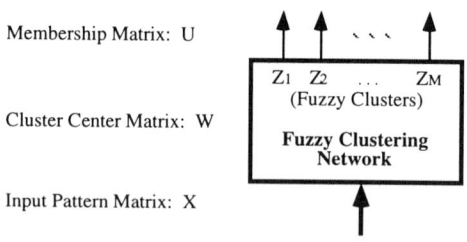

Figure 8.1 Fuzzy clustering network.

8.2 FUZZY CLUSTERING AND NEURAL NETWORKS

In grouping a set of data points in Euclidean space into a given number of clusters, if each point is restricted to belong to exactly one cluster, then a "hard clustering problem" is obtained versus a "fuzzy clustering problem" (where each data point belongs to all clusters with some degree of membership). Historically, the latter problem evolved from the first. Fuzzy clustering should be useful in applications where clusters touch or overlap.

A fuzzy clustering problem can be briefly reviewed with reference to Figure 8.1:

Let $X = \{x_1, x_2, ..., x_P\}$ be an $N \times P$ input pattern matrix. The objective of fuzzy clustering is to generate a partition of X into fuzzy clusters, $Z_i (i = 1, 2, ..., M)$, with membership functions $\mu_{Z_i} : X \to [0, 1]$. Assignment of input patterns to different clusters can be described in terms of an $P \times M$ fuzzy cluster membership matrix $U = [\mu_{ki}]$, where $\mu_{ki} = \mu_{Z_i}(x_k)$, $i = 1, 2, ..., M; k = 1, 2, ..., P$. The element μ_{ki} denotes the degree of possibility that the kth pattern belongs to the ith fuzzy cluster. Also, the elements of U are subject to the following constraints:

$$\mu_{ki} \in [0,1], 1 \leq i \leq M, 1 \leq k \leq P \tag{8.1}$$

$$P > \sum_{k=1}^{P} \mu_{ki} > 0, 1 \leq i \leq M \tag{8.2}$$

$$\sum_{i=1}^{M} \mu_{ki} = 1, 1 \leq k \leq P \tag{8.3}$$

However, each fuzzy cluster center, w_i ($i = 1, 2, ..., M$), can be obtained from an $N \times M$ cluster center matrix, W.

The fuzzy cluster criterion should be generalized for crisp clustering. This leads to many infinite families of fuzzy clustering algorithms and they are formulated into minimization problems, i.e. minimizing the objective function through an iterative method. Ruspini [12] first introduced the problem of fuzzy clustering, in which the objective is to determine the fuzzy classification of each pattern by minimizing some suitably defined functional. Dunn [13] defined the first generalization of the conventional minimum-variance hard clustering. Bezdek [14] generalized Dunn's work into a family of fuzzy clustering problems, developed an algorithm to solve the problem known as the Fuzzy C-Means (FCM) algorithm and gave a comprehensive treatment of the problem [15]. An objective function for the FCM algorithm is given by [13-14]:

$$J = \frac{1}{2} \sum_{i=1}^{M} \sum_{k=1}^{P} (\mu_{ki})^{\beta} (d_{ki})^{2} \qquad (8.4)$$

where $\beta \in (1, \infty)$ is the weighting exponent, and d_{ki} is the Euclidean distance between input pattern, x_k, and cluster center, w_i. To minimize J to find local minimum points which may not be global, its first order optimality conditions are considered to yield the following set of coupled equations:

$$w_i = \frac{\sum_{k=1}^{P} (\mu_{ki})^{\beta} x_k}{\sum_{k=1}^{P} (\mu_{ki})^{\beta}}, \forall i \qquad (8.5)$$

$$\mu_{ki} = \frac{1}{\sum_{l=1}^{M} \left(\frac{d_{ki}}{d_{kl}}\right)^{\frac{2}{(\beta-1)}}}, \text{ for } d_{kl} > 0, \forall k, i, \qquad (8.6)$$

$$if\ d_{kl} = 0\ then\ \mu_{kl} = 1\ and\ \mu_{ki} = 0\ for\ i \neq j \qquad (8.7)$$

The FCM algorithm is based on Equation (8.5), Equation (8.6) and Equation (8.7) as given below.

Initialization. Select membership functions $U^{(1)}$ arbitrarily. Set $\delta = 1$.
Step 1. Compute $W^{(k)}$ using $U^{(k)}$ and Equation (8.5).
Step 2. Compute $U^{(k+1)}$ using $W^{(k)}$, Equation (8.6) and Equation (8.7).
Step 3. If $f(U^{(k)}, U^{(k+1)}) < \epsilon$, stop, where $\epsilon > 0$ is a small scalar and f is some function. Otherwise set $\delta = \delta + 1$ and go to Step 1.

Modified versions of the fuzzy clustering problem were also proposed by some researches. Selim and Kamel [16] explored the mathematical and numerical properties of the FCM algorithm, Kamel, et al. [17-18] generated the fuzzy query processing using clustering techniques and the thresholded FCM algorithm for semi-fuzzy clustering, Pedrycz [19] and Gustafson and Kessel [20] studied fuzzy clustering with partial supervision and covariance matrix, respectively.

Recently, there is a strong direction of linking the fuzzy clustering problem to neural networks. Such neural network models use a densely interconnected network of simple computational nodes. These models are specified by their network topology, node characteristics, and fuzzy competitive rule. The network topology will define how each node connects to other nodes. The node characteristics define the function which combines the various inputs and weights into a single quantity as well as the function which then maps this value to an output. The fuzzy competitive rule specifies an initial set of weights and indicates how weights should be adapted during use to accomplish the task at hand.

Examples of these efforts are the following. Carpenter, Grossberg and Rosen [21] defined a fuzzy ART: fast stable learning and categorization of analog pattern by an adaptive resonance system. Simpson [22] discussed fuzzy min-max neural networks in fuzzy clustering. Dave and Bhaswan [23] presented the fuzzy c-shells clustering algorithm for characterizing and detecting clusters that are hyperellipsoidal shells. Pal and Mitra [24] described a fuzzy neural network based on the multilayer perceptron with back-propagation learning. Jou [25] implemented fuzzy clustering using fuzzy competitive learning networks. Kao and Kuo [26] proposed a fuzzy neural network based on fuzzy classification concept. Newton and Mitra [27] defined an adaptive fuzzy system for clustering arbitrary data patterns.

All these studies mentioned above are based on algorithm level, and their behaviors and characteristics are primarily investigated by simulation on general purpose computers, such as vector computers and workstations. The fundamental drawback of such simulators is that the spatio-temporal parallelism in the processing of information that is inherent to ANN is lost entirely or partly. Another drawback is that the computing time of the simulated network, especially for large associations of nodes (tailored to application-relevant tasks), grows to such orders of magnitude that a speedy acquisition of neural "know-how" is hindered or made impossible. This precludes the actual fuzzy clustering applications in real time.

In the following sections, we will present an FCNN architecture to solve this problem. The approach is to map a fuzzy competitive learning algorithm onto its corresponding fuzzy clustering neural network architecture with on-line learning and parallel implementation. The effectiveness of the FCNN architecture is illustrated by applying it to a number of test data sets, analyzing the hardware complexity and comparing the performance to that of the FCM architecture.

8.3 FUZZY COMPETITIVE LEARNING ALGORITHM

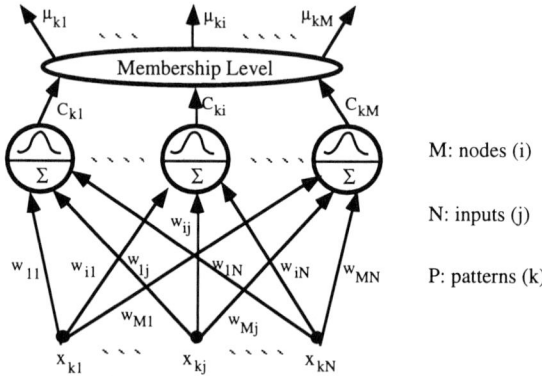

Figure 8.2 Fuzzy clustering neural network model

A basic FCNN model is illustrated in Figure 8.2 [28], where each node represents a fuzzy cluster and the connecting weights from the inputs to a node represent the exemplar of that fuzzy cluster. The square of the Euclidean distance between the input pattern and the exemplar is passed through a Gaussian nonlinearity. The output of the node, therefore, represents the closeness of the input pattern to the exemplar. The degree of possibility that each input pattern belongs to different fuzzy clusters is calculated in the final membership function.

A fuzzy cluster criterion, called *quality of fit or Q*, is defined as the sum of all output values of the nodes over all input patterns, i.e.,

$$Q = \sum_{k=1}^{P} Q_k = \sum_{k=1}^{P} \sum_{i=1}^{M} C_{ki} \tag{8.8}$$

where P and M are the numbers of input patterns and nodes, respectively; C_{ki} is the output of node i when the input pattern is $x_k = (x_{k1}, x_{k2}, ..., x_{kN})$, and

$$C_{ki} = exp\{-\frac{(d_{ki})^2}{2\sigma^2}\} \tag{8.9}$$

The weight vector connecting the inputs to node i is $w_i = (w_{i1}, w_{i2}, ..., w_{iN})$. The Euclidean distance between w_i and x_k is represented as

$$(d_{ki})2 = \sum_{j=1}^{N}(x_{kj} - w_{ij})^2 \quad (8.10)$$

where N is the dimension of the input space. The weight vectors, w_i ($i = 1, 2, ..., M$), can also be viewed as the parameters of the Gaussian functions that determine their locations in the input space. Since the fuzzy clusters are high concentrations of the input patterns in the input space, then locating the weight vectors at or close to the centers of these concentrations will insure a maximum for the quality of fit criterion [29].

This is clearly an optimization problem where the objective function is the *Quality of fit* and the variables are the coordinates of the centers of the Gaussian functions, i.e. the weight vectors. The change in the weight on the objective function (Equation (8.1)) is

$$\triangle w_{ij} = \eta \frac{\partial Q}{\partial w_{ij}} = \eta \frac{\partial}{\partial w_{ij}} \{\sum_{k=1}^{P}\sum_{i=1}^{M} C_{ki}\} \quad (8.11)$$

Which is equal to:

$$\triangle w_{ij} = \eta \sum_{k=1}^{P} \frac{\partial C_{ki}}{\partial w_{ij}} \quad (8.12)$$

where the second summation, $\sum_{i=1}^{M}$, was dropped since w_{ij} appears only one term. Note that

$$\frac{\partial C_{ki}}{\partial w_{ij}} = \frac{\partial C_{ki}}{\partial d_{ki}} \frac{\partial d_{ki}}{\partial w_{ij}} \quad (8.13)$$

and

$$\frac{\partial C_{ki}}{\partial d_{ki}} = \frac{-1}{2\sigma_i^2} C_{ki} \quad (8.14)$$

$$\frac{\partial d_{ki}}{\partial w_{ij}} = -2(x_{kj} - w_{ij}) \quad (8.15)$$

This means that

$$\triangle w_{ij} = \eta \sum_{k=1}^{P} \frac{1}{\sigma^2} C_{ki}(x_{kj} - w_{ij}), \qquad (8.16)$$

where η is a constant of proportionality and σ^2 is the variance of the function of the node. It is clear that this weight adapting utilizes local information available at the weight itself and the node it is connected to. Here, there is no need for an external orientation subsystem to decide which weights are to be increased, since all the weights are adapted on every iteration. The amount of change in the weight is a function of the distance between the input pattern and the weight vector. It is small, for too small or too large distances, so that patterns belonging to different fuzzy clusters do not disturb the weights significantly, and patterns very close to the exemplar will not disturb its convergence.

To introduce a fuzzy competition mechanism between the nodes, a membership function for fuzzy clustering is required. The element μ_{ki} in the fuzzy cluster membership matrix, U, can be represented as

$$\mu_{ki} = \mu_{Z_i}(x_k) = \frac{C_{ki}}{Q_k} \qquad (8.17)$$

It is evident that the elements of U are subject to Equations (8.1 - 8.3).

Using the fuzzy cluster membership element μ_{ki} to participate in the corresponding weight change, a fuzzy competitive learning update rule, which moves the weight vectors towards their respective fuzzy cluster centers, can be represented as

$$\triangle w_{ij} = \eta \sum_{k=1}^{P} \frac{1}{\sigma^2} C_{ki}(x_{kj} - w_{ij})\mu_{ki} \qquad (8.18)$$

We can obtain a variation of this learning algorithm from Equation (8.8). Let

$$\triangle_k w_{ij} = \eta \frac{\partial Q_k}{\partial w_{ij}} \qquad (8.19)$$

Note that the direction of the gradient only guarantees a locally increasing direction. To avoid instability of the algorithm, the step taken in the direction is usually chosen to be very small by the control parameter, η. Thus, if η is

A Parallel ANN Architecture for Fuzzy Clustering

sufficiently small,

$$\triangle w_{ij} \approx \sum_{k=1}^{P} \triangle_k w_{ij} \qquad (8.20)$$

This means that the change in weights will be approximately equal to $\triangle w_{ij}$ if the weights are updated immediately after the presentation of an input pattern. Considering the approximation, Equation (8.18) becomes:

$$\triangle_k w_{ij} = \eta \frac{1}{\sigma^2} C_{ki}(x_{kj} - w_{ij})\mu_{ki} \qquad (8.21)$$

8.4 MAPPING ALGORITHM ONTO ARCHITECTURE

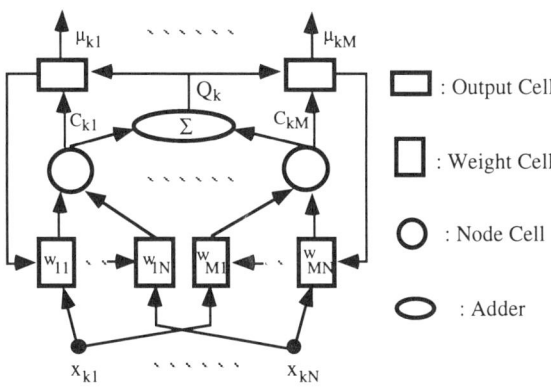

Figure 8.3 The FCNN architecture

From the fuzzy competitive learning algorithm described in Section 8.3, two processing phases, searching and learning, can be defined in FCNN architecture. They are implemented by special feedforward and feedback paths, respectively.

Feedforward Processing:

Two different operations for feedforward processing can be obtained in the algorithm. They are competitive function in Equaton (8.9) and membership function in Equation (8.17). It is evident that the functions can be implemented by the corresponding layers.

Feedback Processing:

The fuzzy competitive learning update rule in Equation (8.18) can be performed by feedback path, since some parameters required for cluster center changing come from the feedforward processing layers. This means that the learning algorithm can be embedded into an architecture.

According to these mapping policies given, a FCNN architecture, with on-line learning and parallel implementation, was obtained and is shown in Figure 8.3.

8.5 FUZZY CLUSTERING NEURAL NETWORK (FCNN) ARCHITECTURE: PROCESSING CELLS

Three kinds of processing cells, i.e. node, output, and weight cell, are used in Figure 8.3 except for an adder ($\sum$) to generate Q_k. A node cell can be implemented by many current approaches [30-33]. Thus, the following discussion will focus on the other two cells, with structures defined in Figure 8.4 and Figure 8.5, respectively. Note that we use a circle as an operator and a square as a memory in both figures.

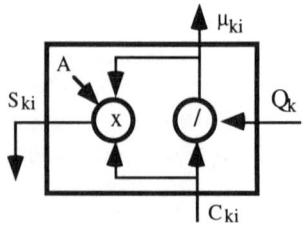

Figure 8.4 Block diagram of the output processing cell

8.5.1 Output Cell Design

The output processing cell is used to obtain both the membership element, μ_{ki}, in Equation (8.17) and the partial product for the fuzzy competitive learning update rule in Equation (8.18), where its two inputs, C_{ki} and Q_k, come from the outputs of node i and the adder ($\sum$) when the input pattern is x_k. The output of the partial product is

$$S_{ki} = AC_{ki}\mu_{ki}, \qquad (8.22)$$

where A is a control parameter for the operator. The output cell can be built by two operators (See Figure 8.4), i.e., multiplier and divider.

8.5.2 Weight Cell Design

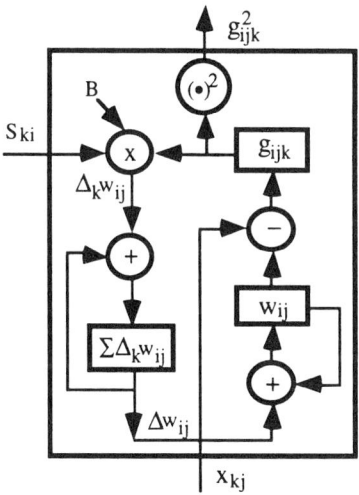

Figure 8.5 Block diagram of the weight processing cell

The weight processing cell is used to store and change weight values, and implement the related arithmetic. It is mainly composed of three memory elements, as shown in Figure 8.5, including accumulation memory ($\sum \triangle_k w_{ij}$), weight memory (w_{ij}) and difference memory (g_{ijk}), and five operators, i.e., adders, subtractor, multiplier and the unit that generates the square of number. In the searching phase, an input, x_{kj}, is given and the output of the cell is represented as

$$g_{ijk}^2 = (x_{kj} - w_{ij})^2. \tag{8.23}$$

In the learning phase, the input S_{ki} is to come from the output cell and its corresponding arithmetic to implement the update rule in Equation (8.21) is

$$\triangle_k w_{ij} = BS_{ki}g_{ijk}, \tag{8.24}$$

where B is the control parameter of the multiplier in the cell, and g_{ijk} is obtained in the previous phase and stored in the difference memory (g_{ijk}). We use the update rule in Equation (8.18) in batch mode, i.e., we accumulate the changes $\triangle_k$ over all input patterns before we actually update the weight, and so the following arithmetic is

$$\triangle w_{ij} = \sum_{k=1}^{P} \triangle_k w_{ij}. \tag{8.25}$$

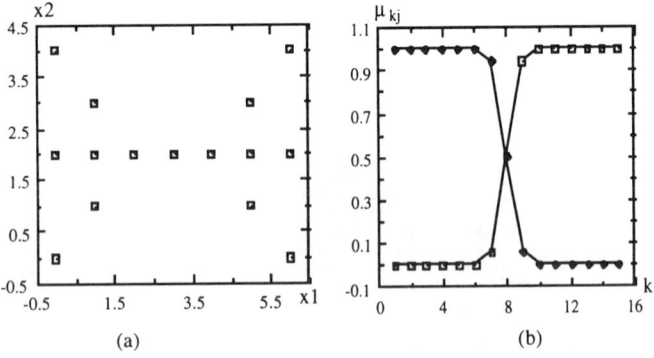

Figure 8.6 The simulation result of applying the FCNN to the Butterfly data set. (a) The Butterfly data input space; (b) The membership value for each point

In this way, we can implement the Euclidean distance between weight vector w_i and input pattern x_k in the FCNN architecture, where Equation (8.10) may be rewritten as

$$(d_{ki})^2 = \sum_{j=1}^{N} g_{ijk}^2, \qquad (8.26)$$

where g_{ijk}^2 is an input from the weight cell w_{ij} to the node cell i. The weights in the FCNN architecture are changed in terms of the following functions:

$$w_{ij}(t+1) = w_{ij}(t) + \triangle w_{ij}, \qquad (8.27)$$

and

$$\triangle w_{ij} = B \sum_{k=1}^{P} S_{ki} g_{ijk} = B \sum_{k=1}^{P} A C_{ki} \mu_{ki} g_{ijk}, \qquad (8.28)$$

where $w_{ij}(t+1)$ and $w_{ij}(t)$ are the weights at time $t+1$ and time t, respectively. Note that Equation (8.28) is a variant of the fuzzy competitive learning update rule in Equation (8.18). We can adjust two control parameters, A and B, equal or close to $1/\sigma^2$ and η, respectively.

8.6 COMPARISON WITH THE FUZZY C-MEAN (FCM) ALGORITHM

8.6.1 Simulation Results

In order to evaluate the effectiveness of the FCNN, we ran many experiments and compared its performance with that of the FCM algorithm. The simulations showed that the clustering results of the FCNN can be similar to ones of the FCM algorithm.

As an example, Table 8.1 shows the results of applying the network to the Butterfly data set [12, 25] with 15 input patterns in R^2. Input patterns, (2, 2), (3, 2) and (4,2), form a bridge between the wings of the butterfly (See Figure 8.6(a)). One interpretation of the butterfly is that patterns in the wings are drawn from two fairly distinct clusters, and patterns in the neck are noise. In the experiment, the control parameters used in the FCNN are $\sigma = 2$ and $\eta = 0.5$, and $\eta = 0.01$ and $\beta = 1.2$ in the FCM. The resulting weights using the FCNN are $w_1 = (5.0496, 1.9999)$ and $w_2 = (0.9506, 2.0000)$. Note that in the FCNN, the membership values, μ_{81} and μ_{82}, are closer to 0.5, leading to a more symmetric clustering than that of the FCM (Figure 8.6(b)). The simulations show that the number of the iterations for the FCNN is also less than that of the FCM algorithm by about the factor of 2.

8.6.2 Hardware Complexity Analysis

Patterns		FCNN		FCM	
k	x_k	μ_{k1}	μ_{k2}	μ_{k1}	μ_{k2}
1	(0,0)	0.0001	0.9999	0.0001	0.9999
2	(0,2)	0.0001	0.9999	0.0000	1.0000
3	(0,4)	0.0001	0.9999	0.0001	0.9999
4	(1,1)	0.0034	0.9966	0.0000	1.0000
5	(1,2)	0.0034	0.9966	0.0000	1.0000
6	(1,3)	0.0034	0.9966	0.0000	1.0000
7	(2,2)	0.0548	0.9452	0.0000	1.0000
8	(3,2)	0.4998	0.5002	0.3492	0.6508
9	(4,2)	0.9451	0.0549	0.9999	0.0001
10	(5,1)	0.9966	0.0034	1.0000	0.0000
11	(5,2)	0.9966	0.0034	1.0000	0.0000
12	(5,3)	0.9966	0.0034	1.0000	0.0000
13	(6,0)	0.9998	0.0002	0.9999	0.0001
14	(6,2)	0.9998	0.0002	1.0000	0.0000
15	(6,4)	0.9998	0.0002	0.9999	0.0001

Table 8.1 Comparison with FCM

Our goal is eventually to implement fuzzy clustering in VLSI. Using the current VLSI technology, the FCNN architecture can be flexibly built by only three types of processing cells plus an adder. It is clear that the cells can be processed in parallel and on-line learning can be obtained. Each cell, which is built from some operators and memories, can be easily implemented based on either the digital or the hybrid approach.

The hardware complexity given by the design is usually considered as an important measure for VLSI implementation. Based on the FCNN architecture, its hardware complexity can be represented as:

$$H_{FCNN} = M(NH_{weight} + H_{output} + H_{node}) + H_{\sum} \quad (8.29)$$

where N and M are the dimensions of the input and output space, respectively; $H_{\sum}$ is the complexity of the adder ($\sum$) in Figure 8.3, and H_{weight}, H_{output} and H_{node} are the complexities of three processing cells, respectively. Note that H_{FCNN} is independent of P, the number of the input patterns, and $H_{\sum}$ has a linear complexity in the number of connections of the nodes, M. As the attached cost of direct competition in the FCNN architecture, $H_{\sum}$ can be compared with the other competitive architecture, such as the MAXNET [34], with the connective complexity, $M(M-1)/2$. This means that the connective cost of direct competition in the FCNN is reduced by a factor of $(M-1)/2$.

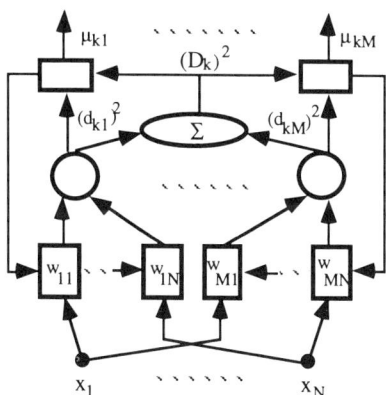

Figure 8.7 Fuzzy c-means neural network (FCMN) architecture

We have shown that the FCNN produced better results in comparison with the FCM algorithm. In order to analyze the effectiveness of the FCNN architecture, we also take the Fuzzy C-Means Network (called FCMN) architecture as our comparative target. An objective function for the FCMN has been given in Equation (8.4). Thus, the fuzzy competitive learning update rule to minimize Equation (8.24) can be obtained in [25], i.e.,

$$\triangle_k w_{ij} = \eta \gamma_\beta (x_{kj} - w_{ij}) \tag{8.30}$$

where

$$\gamma_\beta = (\mu_{ki})^\beta [1 - \beta(\beta - 1)^{-1}(1 - \mu_{ki})] \tag{8.31}$$

with $\beta \in [1, \infty]$.

Like the FCNN architecture in Figure 8.3, the FCMN, with on-line learning, can be implemented using the architecture shown in Figure 8.7. It is evident that these two architectures have identical structure and the same number of building elements, but differ in the complexities of the three different processing cells. The node cell for the FCMN architecture is defined as the Euclidean distance $(d_{ki})^2$ in Equation (8.10) rather than the Gaussian nonlinearity C_{ki} in Equation (8.9). Each output cell is given by

$$\mu_{ki} = \frac{1}{\left(\frac{d_{ki}}{D_k}\right)^{\frac{2}{(\beta-1)}}} \tag{8.32}$$

where $(D_k)^2$ is the output of the adder ($\sum$) and is defined as

$$(D_k)^2 = \sum_{i=1}^{M}(d_{ki})^2 \tag{8.33}$$

The weight cell is represented in both Equation (8.30) and Equation (8.31). Except for the node cell, the other two cells in the FCMN architecture are more difficult to implement by VLSI than the corresponding ones in the FCNN architecture. This is due to the use of more complicated functions, especially, in the weight cell, where nearly five more operators, including multiplier, subtractors and β power exponent, will be needed. However, the Gaussian function defined in the node cell in the FCNN architecture can be, in fact, replaced with two sigmoid functions which are easier to implement in VLSI. One way is to use the difference of the sigmoid functions.

The definitions of the building elements in the two architectures discussed above are summarized in Table 8.2, where the number of the elements required is indicated in parentheses.

8.7 CONCLUSIONS

In this chapter, a parallel FCNN architecture which uses Gaussian functions in the nodes is presented. The FCNN originates from embedding the fuzzy competitive learning algorithm into the neural network architecture. Three types of processing cells plus an adder is developed for the FCNN architecture so that on-line learning and parallel implementation can be achieved. The effectiveness of the FCNN architecture is illustrated by applying it to a number of test data sets, analyzing the hardware complexity of the architecture and comparing the performance to that of the FCMN architecture.

	FCNN	FCMN
Weight Cell (M x N)	1. $g^2_{ijk} = (x_{kj} - w_{ij})^2$ 2. $\Delta w_{ij} = \eta \Sigma S_{ki} g_{ijk}$	1. $g^2_{ijk} = (x_{kj} - w_{ij})^2$ 2. $\Delta w_{ij} = \eta \Sigma (\mu_{ki})^\beta [1-\beta(\beta-1)^{-1} (1-\mu_{ki})] g_{ijk}$
Node Cell (M)	$C_{ki} = \exp\{-\dfrac{\sum_{j=1}^{N} g^2_{ijk}}{2\sigma^2}\}$	$(d_{ki})^2 = \sum_{j=1}^{N} g^2_{ijk}$
Output Cell (M)	1. $\mu_{ki} = \dfrac{C_{ki}}{Q_k}$ 2. $S_{ki} = \dfrac{1}{\sigma^2} C_{ki} \mu_{ki}$	$\mu_{ki} = \dfrac{1}{\left(\dfrac{d_{ki}}{D_k}\right)^{2/(\beta-1)}}$
Adder (1)	$Q_k = \sum_{i=1}^{M} C_{ki}$	$(D_k)^2 = \sum_{i=1}^{M} (d_{ki})^2$

Table 8.2 Comparison of the processing cells in the two architectures

REFERENCES

[1] Kohonen, T., "An introduction to neural computing," Neural Networks 1, 1988, pp. 3-16.

[2] Grossberg, S., Neural networks and natural intelligence, Cambridge, MA: Bradford Book, MIT Press, 1988.

[3] Lisboa, P.G.J., (ed.), Neural networks: current applications, Chapman & Hall, 1992.

[4] Reddy, V., et al., "Survey of neural network architectures," Proc. of the 2nd IASTED Inter. Symp. Expert Systems and Neural Networks. HI, USA, Aug. 1990, pp. 55-88.

[5] Vlontzos, J.A., and Kung, S.Y., "Digital neural network architecture and implementation," VLSI Design of Neural Networks, June, 1990, pp. 205-227.

[6] Mason, R., Robertson, W., and Pincock, D., "An hierarchical VLSI neural network architecture," IEEE Journal of Solid-State Circuits, Vol. 27, No. 1, 1992, pp. 106-108.

[7] Harvey, R.L., DiCaprio, P.N., and Heinemann, K.G., "A neural network architecture for general image recognition," Iincoln Laboratory Journal, Vol. 4, No. 2, 1991, pp. 189-207.

[8] Shim, C., and Cheung, J.Y., Neural network architecture for systolic array, IJCNN-91-Seattle, Vol. 2, 1991, pp. 914-918.

[9] Kulkarni, A.D., et al., "Some applications of parallel distributed processing models," The Workshop on Applied Computing, Stillwater, 1989, pp. 185-192.

[10] Hopfield, J.J., and Tank, D.W., "Neural computation of decisions in optimization problems," Biological Cybernetics, Vol.52, 1985, pp. 141-152.

[11] Hartmann, G., "The closed loop antagonistic network (CLAN)," R. Eckmiller: Advanced neural computers, North-Holland, Amsterdam, 1990, pp. 279-285.

[12] Ruspini, E., "Numerical methods for fuzzy clustering," Inf. Sci., 2, 1970, pp. 319-350.

[13] Dunn, J.C., "A fuzzy relative of the ISODATA process and its use in detecting compact well separated clusters," J.Cybernet. 3, 1974, pp. 32-57.

[14] Bezdek, J.C., Fuzzy mathematics in pattern classification, Ph.D Thesis, Cornell University, Ithaca, NY, 1973.

[15] Bezdek, J.C., "A convergence theorem for the fuzzy ISODATA clustering algorithms," IEEE Trans. Pattern Anal. Machine Intelligence, 2(1), 1980, pp. 1-8.

[16] Selim, S., and Kamel, M., "On the mathematical and numerical properties of the fuzzy c-means algorithm," Fuzzy Set and Systems 49, 1992, pp. 181-191.

[17] Kamel, M., Hadfield, B., and Ismail, M., "Fuzzy query processing using clustering techniques," Information Processing & Management 26, 2, 1990, pp. 279-293.

[18] Kamel, M., and Selim, S.Z., "A thresholded fuzzy c-means algorithm for semi-fuzzy clustering," Pattern Recognition, 24, 9, 1991, pp. 825-833.

[19] Pedrycz, W., "Algorithms of fuzzy clustering with partial supervision," Pattern Recognition Lett., 3, 1985, pp. 13-20.

[20] Gustafson, D.E., and Kessel, W.C., Fuzzy clustering with a fuzzy covariance matrix, Scientific Systems, Inc., Cambridge, MA, 1978.

[21] Carpenter, G., Grossberg, S., and Rosen, D., "Fuzzy ART: fast stable learning and categorization of analog pattern by an adaptive resonance system," Neural Networks, Vol.4, No.6, 1991, pp. 759-772.

[22] Simpson, P., "Fuzzy min-max neural networks – Part 2: Clustering," IEEE Trans. on Fuzzy Systems, Vol.1, No.1, 1993, pp. 32-45.

[23] Dave, N.R., and Bhaswan, K., "Adaptive fuzzy c-shells clustering and detection of ellipses," IEEE Transactions on Neural Networks, Vol. 3, No.5, September, 1992, pp. 643-662.

[24] Pal, S.K., and Mitra, S., "Multilayer perceptron, fuzzy sets, and classification," IEEE Transactions on Neural Networks, Vol. 3, No.5, September, 1992, pp. 683-697.

[25] Jou, C.C., "Fuzzy clustering using fuzzy competitive learning networks," IEEE IJCNN'92, 1992, pp. 714-719.

[26] Kao, C.I., and Kuo, Y.H., "A neural network model based on fuzzy classification concept," IEEE IJCNN'92, 1992, pp. 727-732.

[27] Newton, S.C., and Mitra, S., "An adaptive fuzzy system for control and clustering of arbitrary data patterns," IEEE International Conf. on Fuzzy Systems, 1992, pp. 363-370.

[28] Zhang, D., Kamel, M., and Elmasry, M.I., "Fuzzy Clustering Neural Network (FCNN) using Fuzzy Competitive Learning," World Congress on Neural Networks (WCNN'93), Portland, Regon, Vol.II, July 11-15, 1993, pp. 22-25.

[29] Dajani, A., Kamel, M., and Elmasry, M.I., "A single layer potential function neural network for unsupervised learning," IEEE IJCNN'90, II, 1990, pp. 273-278.

[30] Zhang, D., Jullien, J.A., Miller, W.C., and Swartzlander, E., "Arithmetic for digital neural networks," 10th IEEE Symposium on Computer Arithmetic, Grenoble, France, 1991, pp. 58-63.

[31] Graf, H.P., Jackel, L.D., and Hubbard, W.E., "VLSI implementation of a neural network model," Computer, 1988, pp. 41-49.

[32] White, B.A., and Elmasry, M.I., "The digi-neocognitron: A digital neocognitron neural network model for VLSI," IEEE Transactions on Neural Networks, 3, No.1, January, 1992, pp. 73-85.

[33] Burr, J.B., "Digital neural network implementation," in Neural Networks, Concepts, Applications, and Implementations, Prentice Hall, 1991, pp. 237-285.

[34] Lipmann, R.P., "An introduction to computing with neural nets," IEEE ASSP Magazine, 4, 1987, pp. 4-22.

9

A PIPELINED ANN ARCHITECTURE FOR SPEECH RECOGNITION

Dapeng Zhang, Li Deng and Mohamed I. Elmasry

9.1 INTRODUCTION

Artificial Neural Networks (ANNs) for speech recognition have recently become a subject of great interest. Research into ANNs has not only embraced new pattern classification paradigms and training algorithms using real speech data, but also dealt with corresponding parallel architecture and VLSI implementation which perform the computations required by the algorithms [1].

A variety of ANNs for speech recognition have been developed during the past few years. Numerous studies have demonstrated effectiveness of the multilayer networks with time sequences as inputs to the networks [2]. Several typical examples are: Time-Delay Neural Network (TDNN) proposed by Waibel and Lang, et al. [3-4]; Block-Windowed Neural Network (BWNN) by Sawai [5] and Dynamic Programming Neural Network (DNN) by Sakoe, et al. [6].

A common design approach used in these NNs is incorporation of short delays, of temporal integration, or of recurrent connections. Spectral inputs are sequentially applied to input nodes, one frame at a time, and their corresponding input matrix can be formed. The NNs can thus be integrated into a real time speech recognizer because only short delays are used. However, special architecture which implements such NNs has not been well explored.

VLSI technology offers a highly advanced implementation medium, both at the fabrication and the CAD level, if efficient architecture methods are used. In this chapter, we develop a Pipelined Neural Network (PNN) architecture for speech recognition. The pipelined arithmetic model and its architectural definition are described in Section 9.2. Two processing stages used in the architecture and their building units are discussed in Section 9.3. Case studies for the ANNs and performance analysis for their implementation are given in Section 9.4 and 9.5, respectively. In Section 9.6 we summarize the conclusions of the chapter.

9.2 DEFINITION AND NOTATION

9.2.1 Pipelined Arithmetic Model

A multilayer ANN with time sequence input of feature parameters for speech recognition is composed of $p+2$ layers, which includes an input layer, p hidden layers, and an output layer. Both the input and the hidden layer are characterized by time sequence input matrix of speech parameters, built by $m_s \times n_s$ (s = 1, 2, ..., p+1) memory units, where $m_{p+1} = q$ is the number of pattern classes. The output layer consists of q units. The relation between nodes x_{ij} in Layer s and nodes y_{kl} in Layer $s+1$ can be defined as:

$$y_{kl} = f\left(\sum_{i=k}^{k+e_s-1} \sum_{j=l}^{l+r_s-1} x_{ij} w^{(kl)}_{i-k+1, j-l+1} + \theta^{(kl)} \right) \tag{9.1}$$

where f is a sigmoid function; $w^{(kl)}$ and $\theta^{(kl)}$, $1 \leq k \leq m_{s+1}$ and $1 \leq l \leq n_{s+1}$, are referred to as weight value and bias value from a small input submatrix (called "window") where the size is $e_s \times r_s$ ($e_s \leq m_s$ and $r_s \leq n_s$) in Layer s to the nodes y_{kl} in Layer $s+1$. There will be $m_{s+1} \times n_{s+1}$ windows formed by the input matrix ($m_s \times n_s$) in Layer s.

Based on the window operation model in Equation (9.1), only an input window built by $e_s \times r_s$ units is required in Layer s. Instead of moving such a window to the whole input matrix, speech parameters in time sequence are arranged to pass through the window in pipeline. Thus, the relation in Equation (9.1) can be rewritten as

$$y_{kl} = f\left(\sum W_{kl} X^T \right) = f\left(\sum X W_{kl}^T \right) \tag{9.2}$$

where X is an input submatrix given by such a fixed window, and can be represented as

$$X = \begin{pmatrix} x_1 \\ x_2 \\ \dots \\ x_{e_s} \end{pmatrix} = \begin{pmatrix} x_{11} & x_{12} & \dots & x_{1r_s} \\ x_{21} & x_{22} & \dots & x_{2r_s} \\ \dots & \dots & \dots & \dots \\ x_{e_s,1} & x_{e_s,2} & \dots & x_{e_s,r_s} \end{pmatrix} \quad (9.3)$$

and W^{kl} is the corresponding weight matrix from the window in Layer s to nodes y_{kl} in Layer $s+1$, i.e.

$$W_{kl} = \begin{pmatrix} w_{11}^{(kl)} & w_{12}^{(kl)} & \dots & w_{1r_s}^{(kl)} \\ w_{21}^{(kl)} & w_{22}^{(kl)} & \dots & w_{2r_s}^{(kl)} \\ \dots & \dots & \dots & \dots \\ w_{e_s,1}^{(kl)} & w_{e_s,2}^{(kl)} & \dots & w_{e_s,r_s}^{(kl)} \end{pmatrix} \quad (9.4)$$

It is evident that there are $m_{s+1} \times n_{s+1}$ different weight matrices from Layer s to Layer $s+1$. Their sizes are equal to the size of the window in Layer s.

9.2.2 Pipelined Architecture

The ANNs discussed above can be implemented by a pipelined architecture with $p+1$ processing stages, each with its own sequence of instructions (See Figure 9.1(a)). In each processing stage, a fixed window is built as a connection to the next stage. Loading an input submatrix, X, to the window in a pipeline mode, and mapping the corresponding weight matrix, W_{kl} ($k = 1, 2, ..., m_{s+1}$; $l = 1, 2, ..., n_{s+1}$), the output results, y_{kl}, can be obtained. Such a pipelined neuron structure is given in Figure 9.1(b), where the inputs, $X_i (i = 1, 2, ..., e_s)$, are undelayed or delayed $D_t (= \sum_{j=1}^{t-1} D_j + \triangle$; D is a delay unit and $\triangle$ is its increment, $t = 1, 2, ..., r_s - 1$). The inputs will be multiplied by several weights, one for each delay and one for the undelayed input.

Note that two different types of operations, control flow and data flow, are used in the Pipelined Neural Network (PNN) architecture (See Figure 9.1(a)). The master control unit not only gives each sequence of instructions (control flow) to the corresponding processing stage, but also arranges the parameter of each frame of speech input as data flow input to the given window. Once the window is filled, the submatrix obtained will be processed. Depending on the nature of the speech input, two types of PNN processing stages, using parallel or serial data flow, can be used. These are given in the next section.

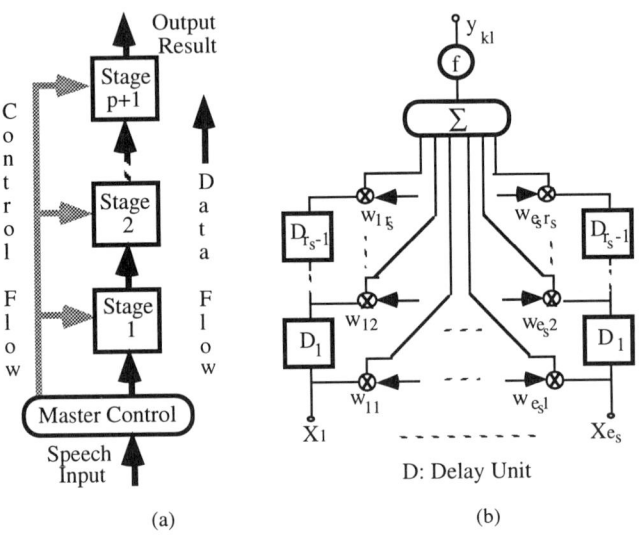

Figure 9.1 (a) Pipelined ANN architecture model; (b) Pipelined neuron unit

9.3 PNN ARCHITECTURE: PROCESSING STAGES

9.3.1 Parallel Data flow Stage

A parallel data flow processing stage in the PNN architecture is shown in Figure 9.2, where a window built by $m_s \times r_s$ shifting units is used to receive the input data flow from its previous stage, and m_{s+1} neurons are used to send the output results to the next stage in parallel. In other words, the input data flow is passed through the window and transformed by this stage to generate the corresponding output data flow. The widths of two data flows are m_s and m_{s+1}, respectively. A new feature parameter obtained by each neuron can be

represented as
$$y_i(s+1) = f(\sum W_i(s) X^T(s)) \qquad (9.5)$$

where $s = 1, 2, ..., p+1$; $i = 1, 2, ..., m_{s+1}$; $W_i(s)$ is the ith weight matrix and $X(s)$ is the input submatrix given by the window in Stage s.

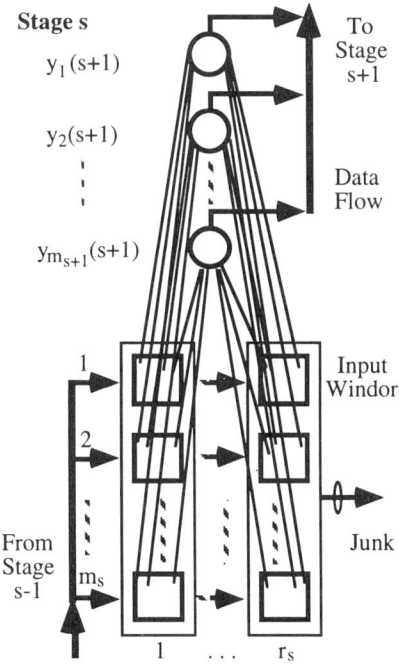

Figure 9.2 Parallel PNN processing stage

9.3.2 Serial Data flow Stage

In this processing stage (see Figure 9.3), a pipe with single parameter width is made by a chain of serial shifting units. A window structure is designed to implement the transformation between stages. The line delays are used to receive a serial stream of parameters and to construct the window required.

The window consists of $e_s \times r_s$ shifting units, which are input to m_{s+1} neurons, and $r_s - 1$ line delays, each $m_s - e_s$ shifting units. The neurons associated with the window are represented in Equation (9.5) with their common output

$$y(s+1) = OR(y_i(s+1)) \tag{9.6}$$

where $i = 1, 2, ..., m_{s+1}$; $s = 0, 1, ..., p+1$. Note that the output of each neuron can be obtained in turn and that there is either a single output, $y_i(s+1)$, or no output within a single clock interval. The output order in a cycle is: $y_1(s+1), ..., y_{m_{s+1}}(s+1), *1, ..., *e_s - 1$, where $m_{s+1} + e_s - 1 = m_s$ and " * " indicates no output. There are a total of $n_{s+1} (= n_s - r_s + 1)$ processing cycles for each given speech input matrix ($m_s \times n_s$).

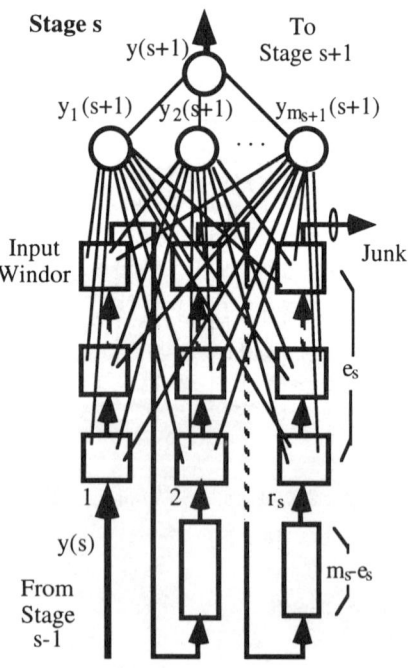

Figure 9.3 Serial PNN processing stage

9.3.3 Building Units Design

Three kinds of building units, i.e. shifting, synapse, and summing unit, are used in each of two different processing stages described above. A shifting unit can be implemented by a regular shifting register and thus the following discussion will be focused on the other two building units. Considering on-line learning, e.g., the BP algorithm [8], which has been successfully applied to the ANN-based in speech recognition, two processing phases, searching and learning, are defined in the units. They are implemented by special feedforward and feedback paths, respectively.

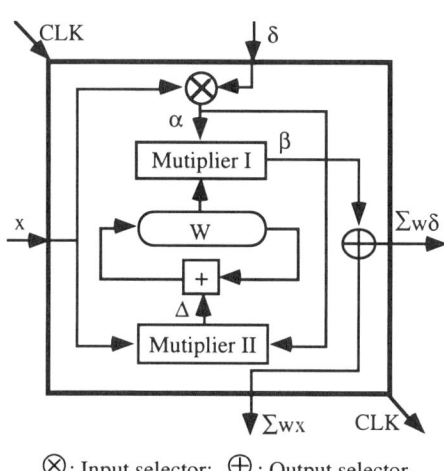

$\otimes$: Input selector; $\oplus$: Output selector

Figure 9.4 Synapse building unit structure

9.3.3.1 The Synapse Unit

The Synapse unit is used to store and change weight values and to implement the related arithmetic. A synapse unit is composed mainly of a weight memory (W), two multipliers (I and II) and two selectors ($\otimes$ and $\oplus$), as shown in Figure 9.4. A control clock, CLK, indicates which phase the unit is, i.e., $CLK = 0$, searching (or feedforward) phase; $CLK = 1$, learning (or feedback) phase. Two data inputs, x and δ, and their outputs, $\sum wx$ and $\sum w\delta$, are given, where x and $\sum wx$ are used for $CLK = 0$, and δ and $\sum w\delta$ for $CLK = 1$. A common Multiplier I is used to generate these two outputs, i.e.,

$$\beta = \alpha W \tag{9.7}$$

where α is the output of the input parameter selector, $\otimes$, and can be represented as

$$\alpha = \begin{cases} x & CLK = 0 \\ \delta & CLK = 1 \end{cases} \tag{9.8}$$

An output parameter selector, $\oplus$, is used to send a correct output result of the unit, i.e.

$$\beta = \begin{cases} \sum wx & CKL = 0 \\ \sum w\delta & CKL = 1 \end{cases} \tag{9.9}$$

Multiplier II is only designed to obtain the increment of the weight value when $CLK = 1$.

$$\Delta = \eta \alpha x \tag{9.10}$$

where η is a gain. Using the arithmetic mechanism attached in the unit, the increment, Δ, can be added to the weight, W, to generate a new weight value.

In this way, when $CLK = 0$, the output of the unit is $\sum wx$ (= wx); Otherwise, the output is $\sum w\delta$ (= $w\delta$), and at the same time, W is changed in terms of the following rule:

$$W = W + \Delta = W + \eta \delta x \tag{9.11}$$

9.3.3.2 The Summing Unit

This unit is used to obtain two-directional accumulative results for both feed-forward and feedback processings (See Figure 9.5). It is built by two amplifiers and a multiplier. Two inputs (outputs), $\sum wx$ and $\sum w\delta$' (δ and y), are from (to) the current stage and the next stage, respectively. Their input / output relations

A Pipelined ANN Architecture for Speech Recognition

are $\sum wx/y$ and $\sum w\delta'/\delta$. When $CLK = 0$, the output of the unit is represented as

$$y = f(\sum wx) \qquad (9.12)$$

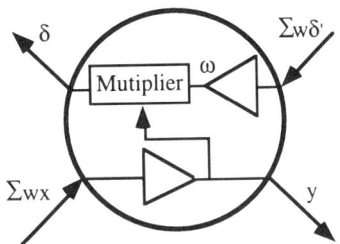

Figure 9.5 Summing building unit structure

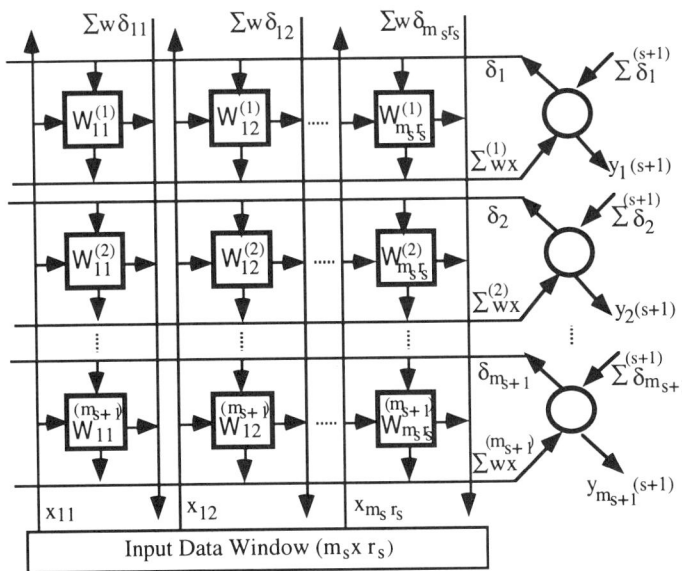

Figure 9.6 Connection network with on-line learning in Stage s

and when $CLK = 1$, the output is

$$\delta = y(1-y)\omega = y(1-y)f(\sum w\delta') \tag{9.13}$$

9.3.3.3 The Connection Network

Using the above building units, a basic connection network in Stage s can be given in Figure 9.6, where the size of the window is defined as $m_s \times r_s$. There are a total of $m_s \times r_s \times m_{s+1}$ synapse units and m_{s+1} summing units used in the network. Clearly, the whole networks can be built by such regular connection networks.

9.4 CASE STUDIES

In this section, we provide the results of our study on three types of neural networks using the PNN architecture. The neural networks selected are motivated by speech recognition applications and they have been widely used by speech recognition researchers [3-6].

9.4.1 Time-Delay Neural Network (TDNN)

TDNN is an architecture that can take into account the "dynamic nature of speech". It is used to represent temporal relationships between successive acoustic frames, while providing some invariance under time translation [3]. The researchers have demonstrated that the TDNN can provide excellent discrimination ability across speech sounds. Speech recognition performance obtained by using the TDNN has often exceeded that of more conventional approaches [4,9].

The basic TDNN architecture is composed of an input layer, two hidden layers and an output layer [3]. Except for the output layer, each layer has a $m_s \times n_s$ ($s = 1, 2, 3$) matrix of memory units, where $n_2 = n_1 - r_1 + 1$, $n_3 = n_2 - r_2 + 1$ and $m_3 = q$. Their relation between the input layer and the 1st hidden layer (and also between 1st and 2nd hidden layers) (See Figure 9.7) is represented as:

$$y_{kl} = f(\sum_{i=1}^{m_s} \sum_{j=l}^{l+r_s-1} x_{ij} w_{i,j-l+1}^{(k)} + \theta^{(k)}) \tag{9.14}$$

A Pipelined ANN Architecture for Speech Recognition   313

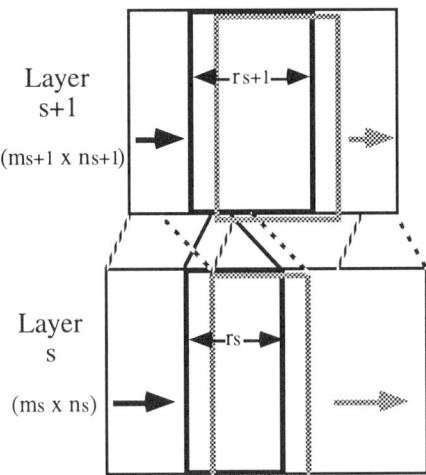

Figure 9.7 Relation between layers for TDNN

The TDNN can be implemented by the PNN architecture with three processing stages which use the parallel data flow structure as shown in Figure 9.2. Except for the last stage, without the data window, each input parameter matrix (m_s x n_s) in the first two stages is pipelined to pass through its window (m_s x r_s), s = 1, 2. When the window is filled by successive data flow, m_{s+1} new values of the parameters are obtained in parallel and simultaneously fed into the window in the next stage. It is evident that there are m_{s+1} different weight matrices and n_{s+1} input data windows from Stage s to Stage $s+1$, and their sizes are equal to the size of the window in Stage s, i.e., m_s x r_s. Using the PNN architecture to implement the TDNN, only m_s x r_s shifting units instead of m_s x n_s (generally, $r_s \ll n_s$) memory units are needed in Stage s.

9.4.2 Block-Windowed Neural Network (BWNN)

BWNN is based on windowing each layer of the ANN with over-laped local time-frequency windows. This architecture makes it possible for the ANN to capture global features from the upper layers as well as precise local features from the lower layers. It has been proved to be robust for speech variations in both frequency- and time-domains among different speakers [5].

The BWNN architecture is composed of an input layer, three hidden layers and an output layer [5]. Except for the output layer, each layer has a $m_s \times n_s$ ($s = 1, 2, 3, 4$) matrix of memory units and their relation satisfies

$$\begin{cases} m_{s+1} = m_s - e_s + 1 \\ n_{s+1} = n_s - r_s + 1 \end{cases} \qquad (9.15)$$

where $m_4 = q$ and $e_s = r_s (1 \leq e_s \leq m_s; 1 \leq r_s \leq n_s)$, the length and the width of the submatrix in Layer s, respectively (See Figure 9.8). Note that the TDNN is a special example of the BWNN if $e_s = m_s$.

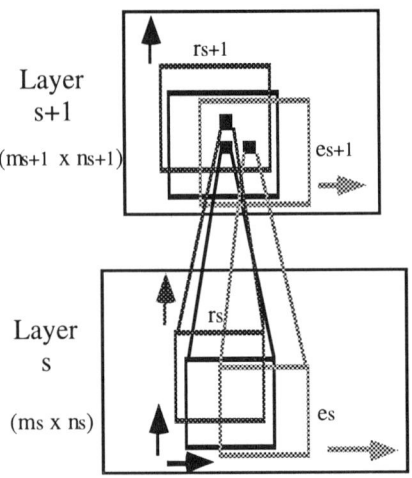

Figure 9.8 Relation between layers for BWNN

Use of the PNN architecture to implement the BWNN involves four processing stages. They are built into the serial data flow structure, shown in Figure 9.3. Like the TDNN implementation, the last stage is as the output stage without the data window. Each input matrix in the other stages can form a $m_s \times n_s$ ($s = 1, 2, 3$) pipeline with the width of a single parameter and passes through its window ($e_s \times r_s$), parameter by parameter. An output result obtained from Stage s within a single clock interval is sent to the window in Stage $s + 1$ without any delay. This means that only an input window built by $e_s \times r_s$ shifting units and some line delays by $(m_s - e_s)(r_s - 1)$ shifting units, i.e., the total $m_s(r_s - 1) + e_s$ shifting units other than $m_s \times n_s$ memory units, are needed in Stage s.

9.4.3 Dynamic Programming Neural Network (DNN)

DNN is proposed on the integration of multilayer ANN and DP (dynamic programming) based matching. Researchers have used DNN extensively in speaker-independent word recognition, and proved that it has excellent time normalization ability, flexible learning facility, expandability to continuous speech recognition, and high tolerance to the spectral pattern variation [6].

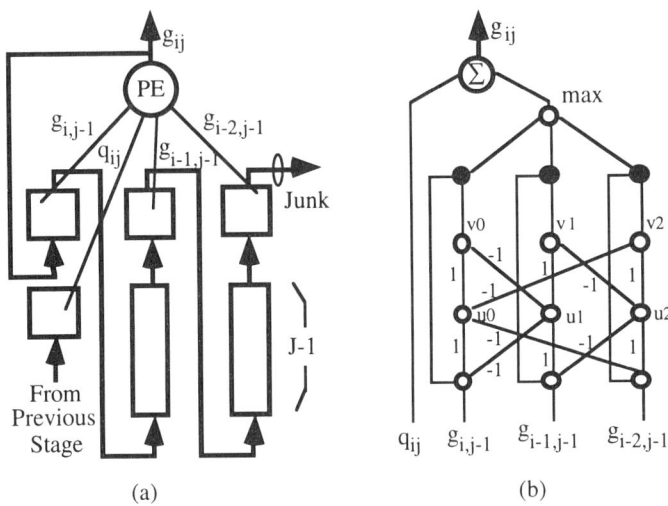

Figure 9.9 (a) Serial processing stage for DNN and (b) its PE structure

The DNN can be implemented by the PNN architecture with three processing stages, where the parallel and serial data flow structure are used in the first two stages and the last stage, respectively. An input pattern, $x_1, ..., x_i, ..., x_I$, is defined as a warping function $i = i(j)$ between input pattern time i and window unit j, where $j = 1, 2, ..., J$. Without an input matrix with $I \times J$ memory units in [6], a window is built by $2x J$ shifting units and J neurons are used in the first stage. When the input patterns x_i and x_{i-1} pass through the window, the corresponding output for each neuron can be represented as

$$y_{ij} = f(w_{j0}x_i + w_{j1}x_{i-1}) \tag{9.16}$$

where w_{j0} and w_{j1} are weighs from two shifting units in window unit j to neuron j. In the second stage, a window with $J x 1$ shifting units is used to receive $y_{ij} (j = 1, 2, ..., J)$, in parallel. Each neuron in the stage is used as a multiplier, i.e.

$$q_{ij} = f(w_j y_{ij}) = w_j y_{ij} \tag{9.17}$$

The third stage is built by a serial data flow structure. Its input data, q_{ij}, is arranged in a pipeline of a single parameter like $q_{iJ} \to ... \to q_{ij} \to ... \to q_{i1}$. In other words, the parallel outputs from the previous stage will be changed as the serial inputs to the this stage. It is composed of a four unit window, two $J - 1$ line delays and a Processing Element (PE), shown in Figure 9.9(a). The PE is designed by the following standard dynamic programming algorithm [10]. Its initial condition is set at $g_{11} = q_{11}$, implemented by the external control. Then, data is processed according to DP-equation.

$$g_{ij} = q_{ij} + max(g_{i,j-1}, g_{i-1,j-1}, g_{i-2,j-1}) \tag{9.18}$$

This maximization problem is implemented by the PE shown in Figure 9.9(b). The PE consists of a tri-comparator subnet for extracting the maximum of three analog inputs [11] and of an adder. Given q_{ij}, its corresponding g_{ij} obtained by the PE is fed into the given window as the input to generate the following new values. This process is continued until the total cumulation value, $Z = g(I, J)$, is reached.

9.5 PERFORMANCE ANALYSIS

To analyze the performances of the PNN architecture, in general, hardware complexity given by the design and access time needed by solving the problem are two important measures [7, 14-15]. In this section, we will analyze these two measures for the PNN architecture discussed in the preceding sections. How to select window parameters for our architecture is also considered in this section.

9.5.1 Hardware Complexity

Based on the current VLSI technologies, the PNN architecture can be flexibly built by two types of processing stages (which we will refer to as *Type*1 and

A Pipelined ANN Architecture for Speech Recognition 317

*Type*2 for parallel data flow stage in Figure 9.2 and serial data flow stage in Figure 9.3, respectively). Using only three simple building units, it is feasible to implement the on-line learning. The PNN architecture can be designed by either the digital or the hybrid approach [12-13].

Based on the case studies in Section 9.4, the PNN architecture can easily implement some typical ANNs used in speech recognition and greatly reduce the memory units in each layer of the ANNs to a small number of shifting units in the processing stage. This is because only a limited window is connected to its next stage and the parameters shifted out from the window are discarded. Since the speech feature parameters are applied to each layer sequentially one frame at a time, this reduction of memory units is easily implementable.

Assuming the number of memory units in Layer s is given as $m_s \times n_s$, and the number of shifting units in $Type1$ is $m_s \times r_s$, a measure of hardware complexity is defined as:

$$H_{Type1(s)} = \frac{m_s \times r_s}{m_s \times n_s} = \frac{r_s}{n_s} \quad (9.19)$$

This means that the hardware complexity, $H_{Type1(s)}$, is related to both r_s and n_s. For the entire networks using $Type1$, the measure is defined as $H_{Type1} = \sum_{s=1}^{p} \frac{r_s}{n_s}$.

Similarly, the number of shifting units in $Type2$ is given as $(e_s \times r_s) + (m_s - e_s)(r_s - 1)$, and its measure is

$$H_{Type2(s)} = \frac{m_s(r_s - 1) + e_s}{m_s \times n_s} \quad (9.20)$$

The measure of the $Type2$ networks, H_{Type2}, is the summation of Equation (9.20) over each stage, e.g., $H_{Type2} = \sum_{s=1}^{p} \frac{m_s(r_s-1)+e_s}{m_s \times n_s}$.

As an example, the basic TDNN architecture for typical speech recognition applications can be described as [3]: $m_1 \times n_1 = 16 \times 15, r_1 = 3; m_2 \times n_2 = 8 \times 13, r_2 = 5; m_3 \times n_3 = 3 \times 9; q = 3$. The number of memory units in the first two layers is $m_1 \times n_1 + m_2 \times n_2 = 344$. However, the number of shifting units of the corresponding stages for $Type1$ in the PNN architecture is $m_1 \times r_1 + m_2 \times r_2 = 88$. Their measure is $H_{Type1} \approx 0.25$, i.e. the hardware complexity of the PNN architecture to implement the TDNN is reduced by a factor of 4.

9.5.2 Throughput Rate

The ANNs for speech recognition are well suited to pipelining because of their multilayer networks as processors of time-sequence patterns. In the PNN architecture embedding parallelism or concurrency, the throughput rate can be fixed and does not vary as the size of the problem grows, i.e.

$$T = O(1) \qquad (9.21)$$

Hence, a high throughput rate can be maintained in such an architecture where the clock of the master control unit is selected from the longest time delay among processing stages.

9.5.3 Window Parameter

The window used in the PNN architecture is an important component and is also an obvious feature which differs from other ANN architectures. The window size has a direct relation to the properties of the PNN, such as the number of shifting units, Ω, and the time delay, Ψ. The smaller the window, the fewer number of shifting units, but the greater length of time delay are required.

In $Type1$, these two trade-off properties for Stage s are

$$\begin{cases} \Omega_{Type1} = r_s \times m_s \\ \Psi_{Type1} = n_s - r_s + 1 \end{cases} \qquad (9.22)$$

We define their product as

$$E(r_s) = \Omega_{Type1} \times \Psi_{Type1} = (r_s \times m_s)(n_s - r_s + 1) \qquad (9.23)$$

To maximize the $E(r_s)$ function, take the derivative with the window size r_s, i.e.

$$E'(r_s) = m_s n_s - 2m_s r_s + m_s \qquad (9.24)$$

A Pipelined ANN Architecture for Speech Recognition

Let $E'(r_s) = 0$, the optimal size of window for $Type1$ can be selected as

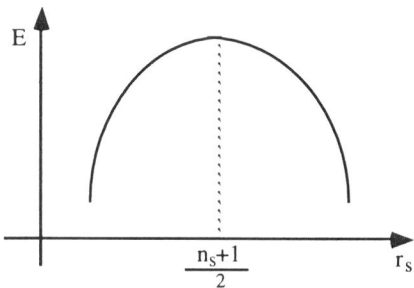

Figure 9.10 Selection of the window for $Type1$

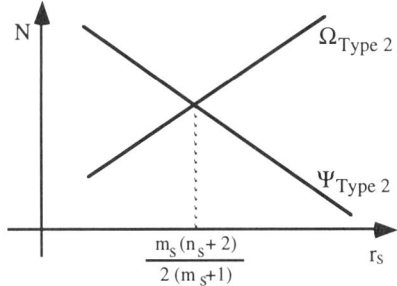

Figure 9.11 Selection of the window for $Type2$

$$r_s = \frac{n_s + 1}{2} \tag{9.25}$$

Note that this optimal window size is not a function of the length of the window, m_s (See Figure 9.10).

In the same way, the two trade-off properties in $Type2$ can be written as:

$$\begin{cases} \Omega_{Type2} = (r_s - 1)m_s + r_s \\ \Psi_{Type2} = m_s n_s - (r_s - 1)m_s - r_s \end{cases} \tag{9.26}$$

where $r_s = e_s$, i.e., a square window is used. The size of the window can be selected directly from the relation $\Omega_{Type2} = \Psi_{Type2}$ (See Figure 9.11), which leads to

$$(r_s - 1)m_s + r_s = m_s n_s - (r_s - 1)m_s - r_s \qquad (9.27)$$

Hence, the choice of the window size for $Type2$ is

$$r_s = \frac{m_s(n_s + 2)}{2(m_s + 1)} \qquad (9.28)$$

9.6 CONCLUSIONS

In this chapter, a novel PNN architecture, which can be used for solving speech recognition problem, is reported. Two efficient processing stages, parallel and serial data flow stages, are developed. We discussed three building units of the architecture: shifting, synapse, and summing unit. As the implementation of the typical ANN models for speech recognition, the PNN architecture can greatly reduce hardware complexity while maintaining a high throughput rate. The performances are analyzed to illustrate effectiveness of the proposed architecture.

REFERENCES

[1] Furui, S., and Sondhi, M.M., (eds.), Advances in Speech Signal Processing, Marcel Dekker Inc., N. Y., 1992.

[2] Lippmann, R.P., "Review of Neural Networks for Speech Recognition," Neural Computing 1, 1989, pp. 1-38.

[3] Waibel, A., Hanazawa, T., Hinton, G., Shikano, K., and Lang, K.J., "Phoneme Recognition Using Time-Delay Neural Networks," IEEE Trans. on ASSP. Vol. 37, 3, 1989, pp. 328-339.

[4] Lang, K.J., and Waibel, A.H., "A Time-Delay Neural Networks Architecture for Isolated Word Recognition," Neural Networks. Vol. 3, 1990, pp. 23-43.

[5] Sawai, H., "Frequency-Time-Shift-Invariant Time-Delay Neural Networks for Robust Continuous Speech Recognition," IEEE, Proceeding of ICASSP-91, S-2.1, 1991, pp. 45-48.

[6] Sakoe, H., Isotani, R., Yoshida, K., Iso, K., and Watanabe, T., "Speaker - Independent Word Recognition Using Dynamic Programming Neural Networks," Speech Recognition, 1989, pp. 439-442.

[7] Brent, R.P., and Goldschlager, L.M., "Some Area-Time Trade-offs for VLSI," SIAM J. Comput., Vol. 11, 4, 1982, pp. 737-747.

[8] Rumelhart, D.E., Hinton, G.E., and Williams, R.J., "Learning Representations by back-propagating errors," Nature, 323, 1986, pp. 533-36.

[9] Waibel, A., "Modular Construction of Time-Delay Neural Networks for Speech Recognition," Neural Computing 1, 1989, pp. 38-46.

[10] Sakoe, H., and Chiba, S., "Dynamic Programming Algorithm Optimization for Spoken Word Recognition," IEEE Trans. on Acoustics, Speech, and Signal Processing, Vol.ASSP-26, 1, 1978, pp. 43-39.

[11] Shih, F.Y., and Moh, J., "Implementing Morphological Operations Using Programmable Neural Networks," Pattern Recognition, Vol. 25, 1, 1992, pp. 89-99.

[12] White, B.A., and Elmasry, M.I., "The Digi-Neocognitron: A Digital Neocognitron Neural Network Model for VLSI," IEEE Transactions on Neural Networks, 3, 1, 1992, pp. 73-85.

[13] Zhang, D., Jullien, G.A., Miller, W.C., and Swartzlander, E., "Arithmetic for Digital Neural Networks," 10th IEEE Symposium on Computer Arithmetic, France, 1991, pp. 58-63.

[14] Burr, J.B., "Digital Neural Network Implementation," in Neural Networks,Concepts, Applications, and Implementations, Prentice Hall, 1991, pp. 237-285.

[15] Graf, H.P., Jackel, L.D., and Hubbard, W.E., "VLSI Implementation of a Neural Network Model," Computer, March, 1988, pp. 41-49.

INDEX

Accuracy and precision, 92, 95
  quantization effects, 103–104, 112
Adaptive Feature Extraction Nearest Neighbor Classifier (AFNN), 251, 268
  architecture, 269
  Discrete Stochastic Complexity for Classification (DSCC), 273
  MMI training, 271
Adaptive Nearest Neighbor Classifier (ANNC), 250, 253
Adaptive Probabilistic Neural Network (APNN), 249, 252
ANNs for Speech Recognition, 199
  Block-Windowed ANN, 313
  dynamic classification, 201
  Dynamic Programming ANN, 315
  hybrid schemes, 203
  Pipelined ANN, 303, 305
  static classification, 201
  Time-Delay ANN, 312
Asynchronous operation, 99, 101, 122, 130–131
Autocorrelation Function, 207, 212
Autocovariance Function, 205–206, 210
Automatic Speech Recognition (ASR), 191, 200
  ASR methodologies, 192
Autoregression, 207

Back-propagation Algorithm, 102, 107, 123, 204, 216, 225
  learning constraint, 108, 123
  local minima, 222
Bayes Rule, 252
Bayesian Inference Framework, 248
  Discrete Stochastic Complexity (DSC), 250, 257
    computational complexity, 260
    Evaluation of, 260
  Discrete Stochastic Complexity for Classification (DSCC), 251
  Hessian matrix, 249
  model evidence, 255
  optimal number of Gaussians, 252
  Probability Density Function (PDF) model selection, 255
  smoothing priors, 248
  stochastic complexity, 255
    asymptotic approximation, 256
  weight Pruning, 249
Biological Neural Networks, 1
Building block circuits, 91, 112, 123
  addition/subtraction, 93
    analog, 94–95, 130
    digital, 94
  data storage, 93, 118
    analog, 94, 97, 117, 131
    digital, 94, 97, 129
    floating gate, 95–96
    multi-functional, 94, 118, 122

multiplication, 93
   analog, 93, 95, 97, 113, 130
   digital, 94–95
  multiplying D to A converters
    (MDACs), 97, 130
  selection of, 98, 114
  thresholding, 93
   analog, 94–95, 131
   digital, 94
Circuit Building Block, 140
  current copy circuit, 147
  error generation circuit, 145
  neuron circuit, 141
  switch circuit, 143
  synapse circuit, 140
Circuit Implementation, 148
  circuit simulation, 150
  Feedforward ANN architecture, 148
  modified Feedforward ANN architecture, 149
  multiple chips, 153
Circuit non-idealities, 92, 95, 97–98
  analysis of, 99, 102
   information theory, 103
   simulation method, 103
   Taylor-series method, 103, 112
  modeling of, 99–100, 105–106, 113, 123
   charge leakage, 114, 118
   Monte-Carlo simulation, 113
Coarticulation, 192, 202, 205
Competitive Learning Networks, 17
  Neocognitron, 19
  self-organizing feature map, 19
  VQ and ART, 17
Concurrency of operations, 97–98, 115, 125
Contrastive Hebbian Rule, 109
Controlling Unit, 51

Correlation Function, 205, 207, 211
Cost function, 112, 116, 120, 125, 128
Data Flow Graph, 92, 99, 101, 111, 118, 123, 126, 128, 130
Design library, 100, 111–112, 114
Design space
  dimensions of, 115
  searching of, 115
Deterministic Boltzmann networks, 97, 109
  learning constraints, 110, 130
  pattern classification, 130
Digi-Neocognitron (DNC) model, 164
Digi-Neocognitron(DNC), 164
  architecture, 165
   C-cell, 175
   S-cell, 171
   Vc-cell, 167
   Vs-cell, 173
  character recognition, 177
  conversion from NC to DNC, 176
  solution, 179
  VLSI implementation, 164
   advantage, 183
Digital ANN
  advantages, 157
  VLSI implementation, 157
Dimensionality Reduction, 250
  feature extraction, 250
Discriminative Training, 249
  Maximum Mutual Information (MMI), 250, 261
Dynamic Programming, 216, 225
Feature Extraction, 268
  adaptive, 268
  Karhunen-Loeve Transform (KLT), 268
Finite State Machine, 51

Index

Forward Algorithm, 195
Forward-Backward Algorithm, 195
Fuzzy C-Means (FCM)
  algorithm, 286
  architecture, 298
Fuzzy Clustering ANN
  hardware complexity analysis, 296
  learning algorithm, 288
  model, 287
  parallel architecture, 292
  processing cell
    output cell design, 293
    weight cell design, 293
Fuzzy Clustering
  butterfly data set, 295
  neural networks, 286
  objective, 285
Gaussian Classifier, 12
Gaussian Mixture, 12
  density estimation, 13, 249
Hardware size, 95
Hidden Markov Model (HMM), 193, 199, 203, 214
  basic structure, 193
  continuous HMM, 195
  discrete HMM, 195
  draw backs, 199
  first order HMM, 194
  Markov assumption, 194
  observation probability, 194
  output independence assumption, 194
  speech modeling units, 197
    microsegments, 198
    phonemes, 192, 197–198, 201
    phonetic feature, 198
    words, 197
  states, 193
  training criteria, 196
    corrective training, 196

Maximum Likelihood (ML), 196, 199
Maximum Mutual Information (MMI), 197
transition probability, 193
High-level synthesis
  analog and mixed analog/digital VLSI, 92, 98
  digital VLSI, 92, 98
  neural VLSI, 98, 114
  scheduling and allocation, 98, 119, 126, 128, 130
    ASAP and ALAP ordering, 119
Hopfield networks, 16, 108
  4-bit A/D converter, 128
  constraint for equilibria, 109, 128
K-Means Algorithm, 13
Layouts, 56–57, 77, 81
Learning Vector Quantization (LVQ), 15, 201, 250
Learning, 93, 108
  Hebbian, 110
  on-chip, 92, 97, 108, 110, 123, 126, 130
  supervised, 110
Linear Discriminant, 14
Markov Chain, 193, 199, 204
Maximum Likelihood (ML) Training, 250, 261
Mean Field Theory, 109
Multi-Layer Perceptron (MLP), 6, 33, 60, 62, 64, 66, 72, 77, 81, 96, 199, 248
  Back-propagation, 7, 248
  character recognition, 64
  classification capability, 6
  conditioning
    Jacobian condition number, 221
    Jacobian rank, 219

digit recognition, 66, 126
issues and limitations, 9
Jacobian, 219
  rank, 219
overfitting, 248
speech prediction, 214–215, 225
speech recognition, 199
XOR problem, 72, 123
Multiplexed mode of operation, 117, 122
Mutual Information, 224, 226
Nearest Neighbor Classifier (NNC), 251
  k-nearest neighbor classifier, 14
Neocognitron (NC), 164
  architecture, 160
    C-cell, 160
    S-cell, 160
    Vc-cell, 162
    Vs-cell, 160
  model, 160
Neural Predictive HMMs (NP-HMMs), 203–204, 214, 216–217
  convergence analysis, 218
  correlation structures, 204
  discriminative training, 223
  joint prediction, 209, 211, 225
  linear prediction, 206
  ML training, 223, 225
  MMI training, 223, 226
  nonlinear prediction, 206
  overtraining, 218
  simulations, 212
  speech recognition accuracy, 218
Neurons, 33–35, 37–40, 42–43, 46, 51, 53, 56–57, 59–60, 62, 64, 66, 68–70, 72–74, 76–77, 79, 81
Nonlinear Time Series Models, 204

Normalized Class Probabilities, 228–229
Optimization procedure, 111
  constraints, 106, 111
  for non-ideality analysis, 99, 105
Partitioning
  netlist, 118
  to analog and digital operations, 98, 118, 120, 123
Pattern Classification, 247
  dynamic, 247
  static, 247
  supervised, 247
  unsupervised, 247
Perceptron, 5
Pipelined ANN
  Architecture
    arithmetic, 304
    parallel processing stage, 307
    serial processing stage, 308
  Building unit, 309
    summing unit, 310
    synapse unit, 309
  Performance analysis
    hardware complexity, 316
    throughput, 318
  Window
    operation, 304
    parameter, 318
Pipelining
  functional, 115, 120–121
Polynomial Classifiers, 12
Probabilistic Neural Network (PNN), 249, 251
Programmable Logic Arrays (PLAs), 51, 54–55
Radial Basis Function (RBF), 10
  learning algorithm, 12
Recognition Systems, 33, 35, 41, 62
  multi-character, 33–34, 60, 62, 66, 70, 79, 81–82

two-character, 33–34, 60, 64, 66, 68, 72–73, 76, 79
Recurrent Neural Networks, 16, 201
Sampled-Data (SD), 33–34, 41
  MDACs, 41, 43, 45–46
  pulse-stream encoding, 43
  Switched-Capacitor (SC), 41, 43, 45–46
  Switched-Resistor (SR), 33, 43, 45, 47, 50, 56, 62, 77, 81
Segmental K-means Algorithm, 216
Simulations, 33
  circuit (HSPICE), 34, 58–59, 72, 79, 81
  software, 34–35, 38–40, 60, 62, 64, 68–70, 72
Speech Recognition Experiments, 214, 226
Sub-threshold operation, 97
Synapses, 33, 40, 46–47, 50, 52, 66, 68, 72–74, 76–77, 81–82
Taylor-series approximation, 99, 103, 118
  computation of partial derivatives, 106, 111
Thresholding, 93, 123
Time Delay Neural Network (TDNN), 16, 202
  NETtalk, 16
Tree-Based Classifiers, 12
Viterbi Algorithm, 195
VLSI, 1, 33
  architectures, 33, 64, 72, 76–77, 82
  implementations, 19, 34–35, 40, 46, 56, 77, 81
    analog, 21, 37
    digital, 21, 40
    mixed analog/digital, 41, 43, 82

parallelism, 20

VLSI ARTIFICIAL NEURAL NETWORKS ENGINEERING

VLSI ARTIFICIAL NEURAL NETWORKS ENGINEERING presents an integrated approach by a single research team to this new engineering discipline. The topics of VLSI implementation of ANNs using analog, digital and sampled data approaches are discussed in detail after an overview of the field of ANNs. Automated VLSI design environment for ANN implementation is covered in depth both at the high and low level of the design hierarchy. Engineering applications of ANNs to character recognition, speech recognition, classification and pattern recognition are also covered. The book can be used as a text or reference for graduate or senior undergraduate courses on the subject. It can also be used by researchers in the field.

M.I. Elmasry is Professor of Electrical and Computer Engineering at the University of Waterloo, Waterloo, Ontario, Canada. He is also the holder of the NSERC/BNR Research Chair in VLSI Design and founding director of the VLSI Research Group. He has published 10 books, over 200 research papers and has been a consultant to many R&D Labs in the area of design and computer-aided design of VLSI digital circuits and systems. He is a Fellow of the IEEE for his contributions to that area.